D1602502

12-22-76

HARVARD EAST ASIAN MONOGRAPHS

68

THE PSYCHOLOGICAL WORLD OF NATSUME SŌSEKI

THE PSYCHOLOGICAL WORLD OF NATSUME SŌSEKI

by Doi Takeo

Translated from the Japanese

Sōseki no Shinteki Sekai

with an introduction and synopses by

William Jefferson Tyler

Published by

East Asian Research Center

Harvard University

Distributed by
Harvard University Press
Cambridge, Massachusetts
and
London, England
1976

The East Asian Research Center at Harvard University administers research projects designed to further scholarly understanding of China, Japan, Korea, Vietnam, Inner Asia, and adjacent areas. These studies have been assisted by grants from the Ford Foundation.

Library of Congress Cataloging in Publication Data

Doi, Takeo, 1920-
The psychological world of Natsume Sōseki = Sōseki no shinteki sekai.

(Harvard East Asian monographs; 68)
Includes bibliographical references and index.
1. Natsume, Sōseki, 1867-1916–Characters. 2. Natsume, Sōseki, 1867-1916–Knowledge–Psychology. I. Title. II. Series.
PL812.A8Z613 895.6'3'4 76-6889
ISBN 0-674-72116-0

This translation is dedicated to

Ichikawa Masaru
(1944–1968)

CONTENTS

PREFACE

A scholar reviewing reports of psychological experiments once commented that German rats appear always to stop and think profound thoughts while American rats run wildly from one part of the maze to another. When psychology goes beyond rats to humans, the cultural component is even more evident. Since psychology as a modern discipline began in the West, it took on the coloration of Western culture. And in no branch of psychology has this been more conspicuous than in psychoanalysis. The very terms found in the analytical literature—Eros and Thanatos, Oedipus and Electra—are redolent with particularities of the West. Yet Western psychology, regarding itself in its grander moments as a branch of modern science, lays claim to universal truth.

Seeking the wellsprings of Western power, most nations of the world have turned to the study of modern science. Since Western science is embedded in the matrix of Western culture, non-Western students have thus been brought into contact with a broad range of Western ideas and social forms. In no non-Western country of the world has this study of the West gone farther and deeper than in Japan.

The author of the present volume, Dr. Takeo Doi, is one of Japan's outstanding psychiatrists. He was formerly Psychiatrist-in-Chief, St. Luke's Hospital, Tokyo; at present he is Professor of Mental Health, Faculty of Medicine, Tokyo University.

As a Japanese psychoanalyst, one of Dr. Doi's central problems is to disentangle universals from particulars in psychoanalytic writings. While using the framework and categories of Freud, Dr. Doi has for years treated them against the actual life experiences of his Japanese patients. In doing so, Doi has wrestled with cultural as well as psychoanalytic assumptions. Some of these assumptions are made explicit in his 1969 article on "Psychoanalysis and Japanese Character." Doi writes that Freud, along with Kierkegaard, Nietzsche, and Marx, in the face of the great social transformations that began in the nineteenth century, variously tried to plan for the restoration of man's human dignity. This, Doi writes, places Freud in the context of Western intellectual history, which cannot simply be transported to Japan as abstract science.

"But in my recent studies of Natsume Sōseki," Doi continues, "I have found a man who, if not the equal of the above giants in intellectual scale and influence, is easily their equal in the sharpness and the depth of his psychological observations. Within the special conditions of his own life, Sōseki was also attempting to restore himself." Born into a modern Japan that was being carried along by currents generated in the West, he experienced their oppression first-hand. He perceived that Western civilization was itself on the brink of bankruptcy. He felt that his personal difficulties were those of all modern Japanese and, indeed, those of all men in the modern world. Consequently, though the psychological patterns that Sōseki depicted were unmistakably Japanese, they were at the same time universal.

To his study of Sōseki Dr. Doi brings a rare combination of intellectual breadth and clinical experience. He may be the most profound and is certainly the most interesting interpreter of Japanese personality to the West. He is fully at home in Western and Japanese cultures. When asked the usual question, "Tell me, why is it that psychoanalysis never flourished in Japan?," he is likely to respond with a question of his own, "That's a fascinating question, but you know, the one country where psychoanalysis has really flourished is America. What is it that is so peculiar about America that explains why psychoanalysis has flourished there?"

Over twenty years ago, Dr. Doi undertook psychiatric training, first at the Menninger Clinic and later at the San Francisco Psychoanalytic Institute. He was a pioneer in the postwar period among Japanese psychiatrists who came to the West for training in psychodynamics. He returned to Japan and began writing books on psychoanalytic theory and practice which at first attracted the interest of only a handful of academically oriented medical and psychological professionals. Later, American scholars interested in the Japanese personality began seeking him out.

Until his book *Amae no kōzō* (published in English as *The Anatomy of Dependency*) became a best seller in 1972, going through more than seventy editions, Doi was hardly known outside narrow academic circles. Now his name and concept of *amae* are likely to pop up as soon as any Japanese begins to talk about Japanese personality. Starting in the 1950s, Doi observed that the Japanese language possessed an intransitive verb *amaeru,* which expresses the primitive desire to be loved and to depend on others. Finding no equivalent in Indo-European tongues, he began to wonder whether the lack of this handy, positively connoted word had diverted the attention of Western psychoanalysts from a study of basic dependency wishes. The concept of *amae* has interested Western readers because, while dependency is universal, in Japan it is marked by peculiar emphases.

Doi, aware that Japanese were reluctant to accept the concepts of psychoanalysis and eager to interest the Japanese public, especially his undergraduate students, happily hit upon the idea of using case material drawn from well-known Japanese literature. He had long been fascinated by the greatest writer of the Meiji period, Natsume Sōseki, and began describing characters in Sōseki stories from a psychological point of view. Having used such examples casually for some years in his lectures in undergraduate courses in psychology, he decided to go through all of Sōseki's works systematically, pulling together a psychological interpretation of Sōseki's major characters. The result is the volume translated here.

Most Westerners undertaking such a study would have attempted to explain the character of the author himself. Somehow the Western intellectual is not likely to think of fictional characters as worthy of psychological studies in and of themselves. Doi is willing to leave aside the question of Sōseki's personality and to accept the reality of the personality types Sōseki so vividly portrays. Doi concludes from his analysis that, just as the penetrating insights of Nietzsche prepared the ground for the psychoanalysis of Freud in the West, so did Sōseki's outpouring of personal agonies in his work prepare the way for the emergence of Japanese psychoanalysis.

The American reader will find this book intriguing for reasons different from those that interest the Japanese. Most of the basic theory of personality expounded in Japanese by Dr. Doi is already a part of America's conventional wisdom. But the American reader will be fascinated to find some of the personality types described in Western literature in Sōseki's very different social setting. Students of psychology will discover the range of these personality types to be rich and rewarding. Students of Japanese literature will discover a vivid introduction to Sōseki's most famous works in a vocabulary and interpretation that open up new levels of understanding. Anyone should find Doi's study a penetrating account of the universal problems faced by individuals coping with a rapidly modernizing society.

William Tyler was a student of Japanese language and culture at International Christian University in Tokyo, 1964–1968, when he first became acquainted with Doi's works on dependency. Later he came to Harvard to do his Masters on the history of Sōseki criticism, and he is presently completing his doctoral dissertation on the contemporary novelist Ishikawa Jun. He also works as a professional interpreter.

He came to know Dr. Doi and, drawn to the book on Sōseki, he decided to translate it with Doi's permission.

January 1976

Albert M. Craig

Ezra F. Vogel

INTRODUCTION

Natsume Sōseki is one of Japan's foremost writers of modern fiction. He was born in Tokyo in 1867 and educated at Tokyo Imperial University where he earned Japan's second bachelor's degree in English literature in 1893. In 1900 he went to London on a government fellowship for further study, returning to Japan a little over two years later to assume a lectureship at his alma mater. In 1907 he abandoned academic life and, in a highly unorthodox move for the times, hired himself out to the *Asahi Shimbun* (The Asahi newspaper) as a writer of serialized novels.

The reading public's first exposure to Sōseki had come two years earlier when, at the insistence of the editor of the influential literary magazine *Hototogisu* (The cuckoo), Sōseki wrote a fifteen-page satiric portrait of "Professor Sneeze," a characteristic intellectual of the day as seen through the eyes of his pet cat. Soon the story blossomed into a popular full-length novel, *Wagahai wa neko de aru* (I am a cat); and it was followed by the novels *Nowaki* (Autumn wind), *Botchan* (Little Master), and *Kusamakura* (The grass pillow). After joining the *Asahi* staff, Sōseki continued to produce a novel a year, in addition to editing a literary review column and writing numerous essays. He gathered about him a coterie of young scholars and writer-aspirants, including the then unknown Akutagawa Ryūnosuke, and through such speeches as "Watakushi no kojinshugi" (My kind of individualism) also established a reputation as a model of character for Meiji and Taishō period youth. In 1916, at age fifty, he died of massive hemorrhaging of his stomach ulcers.

Sōseki depicts in his novels the comic but more often tortured life of the man of intellect. All of his principal characters are university educated and possessed of an overwhelming moral fastidiousness. The broader perspective afforded by their education makes them critical of modern Japanese society, and their high moral standards inhibit them from playing the games of give-and-take that lubricate human relations. Frequently they find themselves feeling alienated from, if not betrayed by, their friends and relatives. At first their alienation assumes the form of the satirical derision of both society and themselves. Later, as their thoughts grow labyrinthine and they become obsessed with the need to achieve genuine rapport with their fellow man, they fall victim to depression, psychosis, and suicide.

In many respects the Sōseki protagonist resembles his author, although with the exception of *Michikusa* (Grass on the wayside) Sōseki's novels are not autobiographical and eschew the "sincere" confessional mode of the *Ich-roman* that dominated Japanese letters in the first thirty years of this century. Despite his many successes, Sōseki was a lonely, unhappy man. An unwanted child, he had been shifted back and forth between his real and foster parents; his marriage was characterized by continual discord; and at a point during or after travel abroad he suffered a serious nervous breakdown; this became the first of several cyclical periods of depression that, along with his writhing gastro-intestinal system, plagued him for the rest of his life. Although he was able to confine his ire to his domestic setting and few outsiders ever suspected him of mental illness, it is known that he was once examined by a psychiatrist; and several pathographs have been written based on the reconstructed facts of his life. Doi would diagnose him as suffering from paranoia.

Doi's interest lies elsewhere, however. His book takes as its province the behavior and attitudes of the central characters that appear in the ten novels written during Sōseki's association with the *Asahi.* Using his professional expertise, Doi brings into relief a number of insights into the human psyche and Japanese society that are latent within the novels. In an introduction to the original edition not reproduced here, he writes, "It is my aim to show how unusual a 'psychologist' Sōseki is. In fact, I never cease to be amazed at his ability to understand the subtleties of different psychopathological states of mind and to write about them in simple, articulate prose. He does not deal in mere psychological description but in thoroughgoing psychological insight; he engages in genuine analytical interpretation born of and refined by his own imagination . . . One could easily call this book 'A Guide to Sōseki's

Psychiatry.'"[1] And, in a conclusion also omitted in this translation, he adds, "Although Sōseki was well read in the field of psychology, there is no evidence that he was acquainted with the writings of Sigmund Freud. His active years as a writer, 1905–1916, correspond to the early stage of Freud's career when Freud was known only in limited circles and his writings were yet to be translated into English, let alone Japanese. Since it is difficult to imagine that Sōseki was influenced by Freud, we are left to conclude that the insight he achieved into his personality through writing novels was, in its brilliance, on a par with that achieved by Freud through analyzing his dreams . . . While Freud's work bore greater fruit in scholastic terms, I am convinced that Sōseki's accomplishments are no less Freud's equal in their depth of insight and universal significance."[2]

In particular, Doi feels that Sōseki observed a psychological phenomenon to which he himself has devoted much clinical attention. This is the subject of *amae,* or dependency wishes. While at no point does Sōseki specifically use the word *amae* with respect to his characters, it was Doi's observation that they invariably suffer from distorted or frustrated dependency wishes. These wishes manifest themselves most openly, for instance, in the characters' obsession with their independence and their refusal to humor or be humored. In Doi's opinion, their personalities, and accordingly the novels, cannot be correctly understood without reference to the psychodynamics of *amae.* (The noun, *amae,* and the verb, *amaeru,*[3] appear in parentheses in the translation wherever they are employed in the original.)[4]

In Doi's paradigm, Sōseki was discontent at being labeled mentally ill and launched upon novel-writing as a means of getting at the roots of his illness. This is not to say that he wrote out of psychosis or simply described his delusions. Rather, drawing upon his psychotic experiences, he created a series of characters who were extensions of his own personality, breathed life into them, and let them act on paper their own private dramas. This required a high degree of objectivity. "This can be accomplished," Sōseki writes, "if one will take his feelings and place them squarely before himself and, stepping back to provide the perspective an outsider would have, conduct a dispassionate and thoroughly honest examination. It is incumbent upon the poet to perform an autopsy on his soul and to report to the world any malaise he may discover."[5] This objectivity did not come easily, of course. It was not until *Sanshirō* in 1908 that Sōseki succeeded in setting the optimum distance between himself and his characters. Prior to this novel, he either lost it in midstream, e.g., *Gubijinsō* (The red poppy); allowed himself to identify with a principal character, e.g., Kiyo in *Botchan;* or to become totally indifferent to his protagonist, e.g., *Kōfu* (The miner). After *Sanshirō,* he succeeded in objectifying his personality through his protagonists, with the technique reaching fruition in the fictionalized autobiography, *Michikusa.* Freed at last from the need to act out his problems in his novels, he turned to writing *Meian* (Light and shade) which, had he lived to complete it, might have qualified according to Doi as the first truly fictional "modern Japanese novel."[6]

From the standpoint of literary criticism Doi's paradigm is not without problems; and the use of characters as alter egos or projections of an author's personality is not confined to Sōseki. Literary critics may also disagree with Doi's methodology and his interpretations, but his acute powers of ratiocination have brought to light previously unnoticed aspects of Sōseki's novels. Principally, Doi is the first to detect the pathological character of Sensei's thinking in the novel *Kokoro* (The human heart); he is the first to offer a satisfactory explanation of the function of the character Keitarō in *Higan sugi made* (By after the equinox); and he is the first to probe the full implications of the triangular love affairs that appear repeatedly throughout the novels.

His book is representative, moreover, of recent thinking in scholarly work on Sōseki. Surprising as it may seem, it is only in the last two decades that Sōseki has come to be treated seriously as a psychological novelist. Although he himself spoke of his indebtedness to the psychological novels of Jane Austen and George Meredith, and his skill at *shinri kaibō* (psychological dissection) was recognized early on,[7] literary criticism in the first forty years after his death was dominated by the view of him as an oriental transcendentalist to the exclusion of almost any other approach. This view was propagated chiefly by Komiya Toyotaka, one of his "disciples" and author of a voluminous biography, *Natsume Sōseki.*[8] According to Komiya, Sōseki's life and works are the record of his struggle with the evil of egoism and of his eventual attainment of an enlightened state called *sokutenkyoshi,* "abandoning self to live in accord with the will of heaven."[9] This was a term that, historically speaking, Sōseki used in the final months of his life to describe a new approach he was applying to the composition of *Meian.* It was not until the mid-fifties when the contemporary critic Etō Jun published his *Natsume Sōseki,*[10] however, that Komiya's view was brought into question. Etō's recent four-volume *Sōseki to sono jidai* (Sōseki and his times),[11] also rectifies errors in Komiya's biography, and it examines the influence of British writers on Sōseki's development as a psychological novelist. Doi's book adds a new dimension to this research, coming as it does from outside the world of literary criticism.

This translation has presented several problems. There is a fundamental difference in the nature of the audience for which it was prepared. Doi could assume that Japanese readers knew a great deal about Sōseki's novels but little of psychoanalysis. The English-speaking reader, on the other hand, will be familiar with psychiatric terms but will know little of Sōseki. Explanations of elementary terminology have been eliminated with the author's cognizance. In addition, a synopsis of each of the novels prefaces each chapter since existing translations of Sōseki's novels vary considerably in quality and the works *Kōfu, Sanshirō, Sorekara* (And then) and *Higan sugi made* are not available in English. The synopses have been written with a view to avoiding interpretation and to giving a straightforward chronological account of the major events in the stories. The reader may also wish to refer to Edwin McClellan's *Two Japanese Novelists: Sōseki and Tōson,*[12] which provides interpretative synopses of seven of the ten novels treated in this book.

I wish to acknowledge the help of Professors Howard Hibbett, Albert Craig, and Ezra Vogel of Harvard University, who contributed greatly to seeing this project through to completion. Many thanks are also due Assistant Professor Jay Rubin and Mr. Gen Itasaka, also of Harvard; as well as my editor, Ms. Florence Trefethen, and my typist, Ms. Judy Treadwell. Above all, I am deeply grateful to Dr. Doi for his helpful supervision and great patience.

Tokyo, 1975 William J. Tyler

Chapter One

BOTCHAN (Little Master)

1906

Synopsis

Botchan, the first-person protagonist, is twenty-three years old. He has graduated from a School of Physics and is about to depart for Shikoku to teach mathematics at Matsuyama Middle School. For one born and raised in Tokyo, this is a remote place indeed.

His parents are dead and an only brother lives far away, but Botchan does not fail to bid farewell to Kiyo, an old family maid. Kiyo has always doted on the boy, although his mischievous, head-strong behavior as a child had been a constant source of embarrassment to his parents. She admires him for his forthrightness and determination—the mark of a true Edokko[1] *and descendant of* hatamoto *samurai*[2]*—and awaits the day when he will install her in a house of his own.*

At Matsuyama Botchan is repulsed by the servility of the local town folk and the pompousness of the school faculty. In a letter to Kiyo, he gives his colleagues nicknames. There is Badger, the headmaster, who resembles the animal; his assistant, Redshirt, who wears a red flannel shirt year round; Clown, an art teacher who dances attendance upon Redshirt; and Porcupine, head of the math department. Although initially Botchan dislikes this strapping man with a bristly head of hair, Porcupine finds him lodgings and treats him to a dish of flavored crushed ice.

It is not long before Botchan too acquires several nicknames. The words, "Professor Tempura," appear on his blackboard after he is seen bolting down four bowls of his favorite noodles with tempura; and the students develop other jokes about a bright red washcloth he uses at the public bath. Finally the ribbing gets out of control one evening when Botchan is assigned to night duty at the school dormitory. While he is off taking a bath, grasshoppers are put into his bed. When he attempts to ferret out the trouble-makers, the students raise such a commotion that the headmaster is called in.

Shortly thereafter Botchan is invited to go fishing by Red-shirt and Clown. As he dozes in the warm sun beating down on the boat, he overhears a conversation between the two men implying that Porcupine is responsible for the grasshopper incident. Angered at what he considers to be a betrayal of his friendship, Botchan decides to confront Porcupine the next day. Before he can say even a word, however, Porcupine announces that Botchan must vacate his lodgings. The landlord has complained to him of Botchan's "rowdy" behavior. Botchan is now convinced of Porcupine's deceitfulness and, as a symbol of his outrage, insists on repaying Porcupine for the dish of crushed ice. Porcupine is miffed by the change that has come over Botchan and will not take the money. Thus the coins are left to sit on Porcupine's desk in the teachers' room.

Meanwhile, a meeting is convened to discuss disciplinary measures for the students. Redshirt, seconded by Clown, favors leniency while Botchan and, much to Botchan's surprise, Porcu-pine are opposed. Porcupine notes, however, that, had his col-league not been briefly and inexcusably absent from the dormitory, the incident would not have occurred.

This prompts a homily from Redshirt who, poking more fun at Botchan's taste for noodles and hot baths, asks that the teachers confine themselves to consolations of a more spiritual nature. Botchan becomes so outraged he cannot resist asking if "the Madonna" qualifies as such a consolation. The name had been bandied about in the boat, and he believes that she is Redshirt's mistress. A deathly silence ensues.

The truth is that the Madonna is the most beautiful woman in Matsuyama and is engaged to a pale, self-effacing member of the faculty named Koga. But Redshirt has decided to take her for him-self and arranges for Koga to be transferred to a school in Kyushu.

When Botchan hears that Koga has no desire to leave Matsu-

yama, he wonders if he too has not been deceived by Redshirt, and he quickly retrieves the coins from Porcupine's desk. Porcupine has learned, meanwhile, that Clown, desirous of Botchan's old accommodations, put the landlord up to evicting Botchan. Botchan and Porcupine mend their differences and vow to take revenge on Redshirt for his treatment of Koga.

Before they can lay plans, a national holiday is declared to celebrate Japan's victory in the Sino-Japanese war. Public entertainment is offered, and Redshirt has his younger brother invite Porcupine and Botchan to attend with a throng of students. A fight breaks out with a rival school and the two teachers jump in to referee. In the process they too receive a pummeling and, to add insult to injury, Porcupine is named as the instigator of the brawl when the story appears in the local paper the next morning. Unable to prove that Redshirt arranged the phony article, he is forced to resign, and Botchan, sacrificing his future as a teacher, joins him in what is now a personal vendetta.

The two rent a room overlooking a local brothel. After eight nights of watching, they catch Redshirt and Clown leaving the premises and, in a comic scene, give them a sound thrashing. Fearful of what the town will say about his extracurricular activities, Redshirt does not dare go to the police.

Botchan returns to Tokyo, sets up house with Kiyo, and finds employment as an engineer. The novel closes as Kiyo is dying. Her deathbed exhortation is to bury her in Botchan's temple.

Botchan's Personality

Botchan, or "Little Master," is known and loved by millions of Japanese readers for his unflinching courage and honesty—for his good character. In the novel, too, he is praised on several occasions by the family maid, Kiyo, for his "good, straightforward disposition," and many readers find that in no time they are sympathizing with him and his rash behavior. Curiously enough, Botchan himself does not consider his personality especially attractive, and he is even suspicious of the praise and affection that Kiyo showers upon him. "I had long ago decided that I was not the type to be liked and it did not bother me in the least to be treated like a block of wood. If anything I found it quite incredible when people would fuss over me as Kiyo did. Sometimes when we were in the kitchen alone she would say what a nice disposition I had. I didn't understand. Were that true why didn't others treat me better? My stock reply was to say that I did not care for flattery, and she would say, 'But that's why you are so nice,' and look at me so happily. It bothered me the way she seemed proud of having created me."

As a child Botchan was, as he says, headstrong, naughty, and given to pranks; and during his brief employment at Matsuyama Middle School he is anything but popular, what with being at cross-purposes with nearly everyone on the faculty. Perhaps there is, therefore, no objective reason to praise his personality. We like him just the same, and I wonder if this is not because Sōseki succeeds in having us look at him through Kiyo's eyes. It may very well be that he is her creation and that she exerts an immeasurable influence on his personality. As a matter of fact, leaving Tokyo makes Botchan realize how irreplaceable she is. We can imagine that she has played a major role in shaping his personality.

The boy was not loved by his parents. "The old man did not pay the slightest attention to me. Mother always sided with my older brother." He lacked any opportunity to indulge in his parents' affection (*amaeru*); nor did he try to create one. He was an ugly duckling. Into this situation came the uncommon love Kiyo bore for him. At first he resisted it. "In my childish mind I questioned why she should love me. 'It's nonsense. If only she'd stop it,' I thought. I felt sorry for her." There are instances, too, of his refusing to accept presents that she had bought without his parents' consent on the grounds that such partiality was unfair to his brother. In other words, Botchan would not let himself be spoiled (*amaeru*) by either his parents or Kiyo. This is not to say he disliked Kiyo. As time passes a bond develops between them, and Kiyo becomes the most important person in his life.

Why is she so fond of him? According to what Kiyo tells Botchan, she felt sorry for him because he received so little affection from his parents. According to him, however, he was not as wretched as she would have us believe. "It was only because she was constantly finding a pretext to say how unfortunate and unhappy I was that the thought ever entered my head." Are we to assume then that Botchan was merely annoyed at Kiyo's arbitrary decision to treat him as a person in need of pity? I think not, for it is undeniably clear from this passage that he has acquiesced in Kiyo's view of his fate. Undoubtedly Kiyo had lived an unrewarding life and she could appreciate his plight better than he.

There are passages, moreover, that suggest that she derives pleasure from doting on him. In general, whenever a person alternates between feelings of compassion and affection for another, he is engaging in the psychodynamics of obtaining self-gratification through identifying with and bestowing affection on the other. This is apparently the way Kiyo thinks. "She had made up her mind about living with me after I had established my own home. She asked for a position any number of times. I promised, for what it was worth, because I imagined matters would eventually work out that way." When Botchan leaves for his new position he does not cry, but the tears well up in Kiyo's eyes and she hates to see him go. Again, her deathbed wish is "to be buried at Botchan's temple. I shall look forward in my grave to the day when you come to join me." Undoubtedly pity initially draws Kiyo to Botchan's side. But one is forced to conclude that it is she who in the final analysis needs love and affection more. She exerts her utmost for her "Little Master," but she also secretly wishes to be taken care of (*amaeru*) by him.

So far we have examined the relationship in terms of Kiyo, but what significance does it hold for Botchan? There is little doubt in my mind that it has a decisive effect on his personality. Of course, his personality had already been partially formed before she came to work for his family. He had been an unmanageable child, and the cause lay in the troubled relationship with his parents. He grew up a person "who had no desire to use charm or flattery" and who could not compromise readily. But because of Kiyo's love—or rather because of his ability finally to accept it—he learns to relate to people who will wait and dote upon him in a similar manner. He secretly enjoys such treatment, although he is unable to reciprocate in kind. People who will not serve him, consequently, represent potential enemies. Put another way, Botchan is the type of person referred to by expressions such as "a grandmother's boy" (*obā-san ko*)[3] or "an overprotected child" (*kahogoji*).[4] Once a "Little Master," always a "Little Master."

Botchan's immaturity emerges shortly after his arrival in Matsuyama. He tips excessively when he feels poorly treated at the inn and mistakes the friendly smiles of the maids as derision. At school, too, he feels hostility emanating from the headmaster and faculty; he looks upon his classroom as enemy territory. So it is that in no time he has trouble with his students. Admittedly, students will tease a new teacher and, as he states at the faculty meeting, the fault lies with them. We must not forget, however, that not all new teachers are the butt of students' jokes. To a student there is a type of teacher who asks for trouble. Isn't Botchan teased because his pupils immediately perceive something childish about him? Hadn't he once been a prankster himself? We can interpret their pranks as an expression of their sense of secret affinity with him.

A detailed analysis would probably uncover the decidedly neurotic and unattractive aspects of Botchan's personality. But to write him off so easily would hardly be fair. For despite the fact that he cannot readily play up (*amaeru*) to others, and despises those who can, he is blessed with a perceptive and critical spirit in a world where social and moral indulgence (*amae*) is an accepted way of life. This spirit is the source of his superior moral sensitivity, and it accounts for his hatred of Redshirt and his sycophantic crony, Clown. People who use kind words and deeds to take in others for their own selfish ends are, to Botchan's way of thinking, villains and hypocrites.

The reason we derive great satisfaction from his actions is that we are secretly in agreement with him. Whether it be Meiji Japan, when the novel was written, or contemporary Japan, society is rife with characters like Redshirt and Clown who are attuned to the times and profit by their awareness. Likewise, the number of people who have been duped by such men and who belong to the class of wan Kogas is by no means small. That is why we feel for a moment spiritually renewed by a Botchan who is determined to go his own way even if it means his own downfall.

Living with One's Debts: on ni kiru[5]

Falling for Redshirt's slanderous remarks, Botchan begins to suspect that Porcupine is behind the grasshopper incident. He recalls with chagrin the dish of flavored ice and tries to force him to take back the few pennies for the treat. He explains why:

"When I arrived here, it was Porcupine who was the very first to treat me. But to have been treated even to a mere flavored ice by such a two-faced type is a blot upon my honor. As I had only one glass his expense could not have been more than a sen and a half, but, regardless of how trivial the cost, I will never feel content being indebted to a swindler. The money gets returned first thing in the morning.

"I once borrowed three yen from Kiyo. That was five years ago, and I have yet to repay her. It's not that I can't; I just don't. Kiyo is not relying on the money, and the thought never enters her mind to ask herself when it will be returned. As for myself, I have no intention of acting like a stranger who, conscientious about his debts, keeps promising to make payment. That would be as though I were questioning the spirit in which the money was lent—or suggesting that there was something less than beautiful about Kiyo's motives. Do not mistake me. It is not that I feel I can walk all over her. I think of Kiyo as a part of myself.

"Between Kiyo and Porcupine there never will be any comparison. But to accept a favor from an outsider—whether it be an ice or a cup of tea—and to make no move to return it is a way of indicating one's respect for him and the high opinion in which one holds his character. All this fuss could be avoided by having each person pay his own way, but to have a man retain in his heart a feeling of being genuinely indebted (*on ni kiru*) for a favor received—that is repayment that goes far beyond monetary considerations. I may not be a person of importance, but I am my own master. Surely the bowed head of the independent man is a 'thank you' worth more than millions."

This is an extremely interesting and revealing soliloquy. But one is also tempted to say it does not tell the whole truth. Let us consider for a moment the premise that the silent and inward acknowledgment of indebtedness constitutes the best possible means of thanking another and demonstrating one's favor toward him. Botchan says that it is out of respect for the integrity of one's benefactor that the independent man bows his head and makes himself subordinate. This is a truly profound statement: it touches probably on the deepest ethical meaning of the act of gratitude. But in this story Botchan uses this lofty insight to legitimatize his own actions. Without respect there is no need to feel indebted. He had once respected Porcupine, but now that he has learned of his unworthiness he can no longer tolerate being in his debt and returns the cost of the treat. On the other hand, he need not repay Kiyo immediately because she qualifies as the type of person to whom he may be justly indebted.

So far his syllogistic logic is exceedingly clear and, at first glance, seemingly plausible. Examined carefully, however, I believe it will not explain his actions. Botchan can no longer respect Porcupine because he really believes Redshirt. But does not the blanket acceptance of gossip constitute an act of disrespect? I am not convinced, moreover, that his nonchalance about his debts to Kiyo constitutes respect either. He may respect her integrity, but wouldn't we be closer to the truth to say that his nonchalance stems from his view of her as an extension of himself? Kiyo is no more an independent individual for Botchan than he is for her. The two seek to be one. I have already stated that Kiyo identified with Botchan. The reverse is also true.

In other words, while it appears that on a verbal level Botchan possesses a profound understanding of the nature of gratitude, in real life he chooses as binding only those relationships in which both parties seek to become one or, to put it more precisely, in which both believe they are one. But it is extremely easy to feel indebted in a relationship in which what is due is never repaid because neither party is a truly independent person. Both are mutually dependent (*amaeatta*) upon each other. I said that Botchan could not play up (*amaeru*) to people, including Kiyo. That is, at least, what his outward behavior tells us. Clearly the desire to be loved (*amae*) is repressed in him. What is equally clear, however, is its disguised manifestation in his attitude toward Kiyo and, in particular, in the pleasure he derives from being indebted to her.

We can imagine that his secret dependency wishes (*amae*) are directed not only to Kiyo but, in fact, to Porcupine and all the other characters. The sole difference is that, whereas these wishes receive their quiet gratification from her, in every other instance they end in frustration. This difference helps us to understand the full significance of the sense of compulsion with which he returns Porcupine's money. He is not making a simple objective judgment when he says that Porcupine no longer merits respect. He feels betrayed, for, as he says, "to have been treated even to a mere flavored ice by such a two-faced type is a blot upon my honor." Obviously, Botchan believed Porcupine was his ally. He secretly believed that, like Kiyo, Porcupine would stand by him and think only of his welfare. This belief reveals the extreme naïveté of his dependency wishes (*amae*); and the feeling that he had been betrayed at his most vulnerable point accounts for the extremity of his subsequent behavior. Only wiping the slate clean of any and all debts will make him feel right again.

So far we have considered Botchan's attitude concerning indebtedness. While most readers will find it a bit extreme, none will think it abnormal. We can all appreciate his feelings. There is nothing unique about Botchan because he personifies attitudes common to all Japanese. When we say thank you we are apt to feel that one day the favor will have to be returned or otherwise carried to the grave. This is not an altogether pleasant feeling and, examined closely, it does in a sense delimit our freedom. We find it confining because the relationship of oneness such as existed between Kiyo and Botchan is a rarity in human relations: the vast majority of our relationships are fraught with conflicts over self-interests. Even in filial and marital relations, self-interests clash; and it is often the case that the freedom and independence of the persons involved are not at all guaranteed. Thus, on the one hand, we appreciate a kindness while, on the other, we fear to be bound by it. As a matter of fact, I doubt whether any relationship of oneness ever totally escapes the problem of self-interest; but Botchan is at least saved by his conviction that he and Kiyo are of one mind.

We cannot help feeling a little envy for Botchan because he succeeds in achieving the unique relationship of being indebted but not confined. We ache with the restrictions inherent in day-to-day interpersonal relations, and try as we might we can never escape them. We are moved to cheer Botchan in his dogged refusal to become involved in any association that threatens to confine him. We can imagine the source of this novel's popularity lies in the fact that he realizes a dream carried in the hearts of all of us.

Chapter Two

KŌFU (The Miner)

1908

Synopsis

This novel concerns a traumatic week in the life of a nineteen-year-old as he recalls it several years later.

A family crisis over the boy's refusal to marry a girl of his parents' choosing precipitates his leaving home. They want him to marry Tsuyako, but he is in love with Sumi'e. As the story opens, he is weary from walking all night and is about to rest at a teastand on the outskirts of Matsubara. A later passage suggests that he is headed for Kegon Falls, a common spot for suicide in the Meiji era. He has contemplated suicide before but had never been able to summon the courage.

Already seated at the stand is Chōzō, a crimp for the copper mines. His cold, scrutinizing eye sees in the boy a potential miner.

Although the boy is repulsed by the man's filthy appearance and coarse language, he joins him in a cup of tea. Chōzō talks of the money to be made as a miner, but the boy hardly needs conning. No work could be better suited to one who wants to bury himself. He entrusts his fate to Chōzō and follows him docilely back to Matsubara to catch the train.

He is amazed at his own passivity. He considers the possibility that Chōzō is a swindler and announces that he can get to the mine himself. But he has no idea of directions and is embarrassed when Chōzō asks him if he has money for a ticket. He had not thought of the necessity of purchasing one, and there is little money left in the elegant alligator-skin purse he hands Chōzō. Chōzō buys the tickets and pockets the wallet.

Alighting from the train, the two men begin an overland trek to the mine. The surroundings seem unreal to the boy, but anxiety takes second place to hunger. He stares longingly at the eating houses along the road hoping that Chōzō will feed him, but Chōzō is too busy looking for more recruits. He spies another boy wrapped in a red blanket.

The party trudges on. As the sun sets, the boy, dressed in a thin summer kimono, shivers in the chilly mountain air. Presently they come upon a thirteen-year-old ragamuffin whom Chōzō also propositions. The night is spent in a mountain shanty. The next morning they cross a mountain pass in a cold drizzling rain and enter a garish boomtown. Ten thousand men labor in the mine and live in row after row of large, gray dormitories.

Chōzō introduces the boy to the mess hall boss of one dormitory and disappears with the other two recruits. The mess hall boss is a kindly man named Hara Komakichi. This is not the first time he has seen a runaway from a good home, and he urges the boy to return to his family. The boy refuses.

He is led upstairs to a large room. Surrounding a hearth is a group of black, savage faces. None of the miners will let him near the fire, and they deride him for thinking he can become one of them. He is served a dinner of hard, tasteless rice that he is barely able to swallow. Suddenly there is a great commotion outside. The miners rush to the windows shouting as a funeral procession passes through the grounds.

After a restless night of scratching caused by bedbugs, the boy is taken on a tour of the mine by a guide. The cold, dripping walls cause their oil lamps to sputter; the tunnels reverberate with the sound of hammers and blasting; and, as the passages narrow, the two men crawl on all fours. They make a harrowing descent down a series of rope ladders to the very bottom of the mine and, as they are climbing out, the boy becomes dizzy and almost loses his grip. Death lurks everywhere, but this is not the death he envisioned for himself.

At this point he loses his guide. Groping in the dark, he comes upon a powerfully built man. He is struck by the gentility of the man's speech.

The man's name is Yasu. Like the boy he had been a higher school student[6] *but, at age twenty-three, he committed a "grievous crime against society" as a result of a liaison with a certain woman. He has hidden in the mine these last six years, but while the statute of limitations on the crime is to expire soon, he cannot forgive himself. Being a miner has become his destiny.*

Here in the bowels of the earth, the boy finds a kindred spirit. Yasu encourages him to return home.

The boy is medically disqualified for a job as a miner. Unable to leave his newly found friend, he stays and works in the mess hall. Five months later he returns to Tokyo.

In the closing sentence the reader is reminded that the story is a factual account. Sōseki writes that it is based on the life of an anonymous young man he interviewed at his Asahi Shimbun *office in the fall of 1906.*

The Fugitive

The novel begins with a description of the central character. He has run away from home. Except for a nap, he has been walking since he left Tokyo at nine o'clock the night before, and he is exhausted. There is no direction to his steps. Only the force of his decision to leave home drives him on. "If I stopped to rest, they would catch me for sure."

He does not believe that walking will lead him out of his present anxiety. On the contrary, "the more I walked the more I seemed to sink deeper and deeper into a haze from which there was no exit . . . The darker it became, the better. From the dark into the darker. Before long the whole world would be enshrouded and I would no longer see my own flesh. How good that would feel." It is not enough to have left home; he wishes to escape to a world where there is no need to worry about what people think or what he himself thinks. "Anywhere. As long as no one is present and I can live alone." He has embarked on an aimless journey because he does not have the courage to take his own life. If it fails, there will be time enough for suicide. This ultimate step appears to be his sole consolation.

The boy is the son of a person of considerable social standing. He has recently turned nineteen and taken such a liking to a certain young girl that his personality literally would become malleable in her presence. "She would become round and soft, or hard and squarish, in her attitude toward me. When she did, I too became round or square." In the meantime, his parents decided upon another girl as his future spouse. He felt guilty at the way he was spoiling this second girl's chances for a happy marriage, and he made a deliberate effort to fall in love with her. His efforts proved unsuccessful, however; and his mounting sense of guilt served only to remind him of his enslavement to the first girl. This guilt was quickly detected by his parents who, jumping to unwarranted conclusions, refused to believe him when he denied any illicit behavior. So that while, on the one hand, he felt indignant at not being trusted, on the other, he feared that the situation might well lead to the sort of relationship of which he was being accused. And how embarrassing that would be for the second girl! "I was attacked on all sides by utterly incompatible feelings. There was no relief for my tortured brain." When he realizes that "I am the one who is suffering and I alone can stop the pain," he decides to "suddenly 'go up in smoke' and disappear from the mix-up. What other means was there but suicide? I set about the preparations on several occasions." At the moment of truth, however, he loses his nerve. He shelves the idea and leaves home.

It is clear that the boy is driven to these thoughts because he is caught in the "dilemma of *giri*[7] and *ninjō*,"[8] the conflict between social obligation and personal feelings. The phrase sounds old-fashioned to modern ears, and many Japanese will think it has nothing to do with them. This is a serious mistake for, with the exception of the manner in which it presents itself, the dilemma is still very much with us. What we ought to note is that attitudes toward it have changed considerably. In the past people were compelled to choose between either one or the other. In most cases they bowed to the iron will of duty; only in rare instances did they summon enough courage to insist on their own feelings. People today, on the other hand, dislike as a matter of principle being forced to make either-or decisions and prefer a unification of alternative choices. It is characteristic of modern man that he chooses not this or that but the best of both.

In this sense the central character of *Kōfu* has a contemporary bent to his personality because he attempts to reconcile contradictory demands. "I tried everything. I toyed with every possibility until I did not care anymore." In the end he resigns himself to the fact "that the situation will not gel as I hoped." He discovers that his frustration is his and his alone, and he decides to destroy himself.

It is very important to note that it is in the extremity of trying to annihilate his self that he awakens to its very existence. Certainly an awareness of one's self as an individual is born of friction with one's surroundings, but friction alone will not suffice. For invariably the surroundings are blamed for the discomfiture, and it is the surroundings that are expected to devise a means of ameliorating it. The character in this novel says that at first he blindly relied on the outside world to alleviate his pain. "I told myself that, if I could put someone to work for me, then everything would turn out all right. It was as though I had met a stranger on a narrow path and I stood there dumbly hoping that he would step into the mire to make room for me. I proposed the impossible idea that I should stand where I was and he should do as I wished." This surely can be called self-assertion, but it hardly signifies the presence of an independent self and the boy never progresses beyond the egocentric desire to depend upon others. One suspects, moreover, that the "this and that" eclecticism and the so-called individualism of many contemporary men and women consists of this type of camouflaged egoism. These are the people who curse the yoke of family obligations and who censure society as feudalistic. They are in pursuit of more meaningful ties with their fellow men. Of course such ties are clearly essential in any age, and one can legitimately expect that they will assume a form in keeping with the times. But they can never be genuinely achieved if built upon the selfish desire to depend upon others.

When a man cannot find his salvation in his surroundings, he is left with himself. Such is the case of this boy: he confronts his naked self alone. But the birth of one's awareness as an individual is a moment fraught with danger. Simultaneously with the moment that the boy recognizes that he is alone in his suffering, he administers a *coup de grâce* to his own self. He cannot bear to continue living as a totally independent individual.

There is nothing unique about his experience. Probably any number of people would find it intolerable. For many moderns, life as a mere individual is a meaningless existence and, when the social reform ideologies upon which they pin their hopes prove illusory, they are in danger of falling into the same predicament as this boy. Or they may deceive themselves into finding salvation in some form of existentialism, although existentialism by its very nature precludes salvation. In the end when hope is gone, they will attempt to snuff out their lives or, if that is impossible, wander aimlessly out of an uncontrollable restlessness. This is, of course, the realm of neurosis and psychosis. I cannot help but think that the present increase in the incidence of these illnesses is due to the inability of more and more people to come to terms with themselves as individuals.

Encounter

No longer able to contain his feelings, the boy leaves home and starts walking. He presently comes upon a roadside stand, where he finds a nondescript individual staring intently at him. The stare is so penetrating, in fact, that he decides against stopping. Suddenly the man calls out and, as suddenly, the boy turns and walks toward him. His initial feelings of intense dislike "disappeared. It was with a strange feeling of warmth that I retraced my steps."

What is happening here? The boy can hardly explain the contradiction to himself. Hadn't he been seeking death and a place to die? He understands only that the impulse to respond to the man is a "happy one" and concludes that "there is absolutely nothing consistent about human nature." These ambivalent feelings are, however, an extremely common characteristic of human behavior.

An ambivalent person often unconsciously desires the opposite of what he says; and, in this case, we find the boy secretly harbors the desire to cling to people although he claims to be thoroughly disaffected with them. It happens that these feelings come to the fore in this encounter.

But why in this particular instance? The man, Chōzō, makes his living shanghaiing men for the mines. He means no kindness. Yet the boy not only fails to question his motives but also blithely accepts his offer of employment. One reason may have been that he wanted to debase himself, but it hardly explains the warmth he experiences at meeting Chōzō. Only when they arrive at the train station does he begin to think that Chōzō is being too kind and that perhaps there is an ulterior motive to his kindness. At that, it takes Chōzō's asking if he has money for the train to arouse the boy's suspicions. In retrospect he realizes that "in the back of my mind I had the strange feeling that if I stuck with Chōzō he would take good care of me."

Why is it that he experiences this sense of rapport and simple trust? Isn't it because Chōzō is a complete stranger? His family has been an utter disappointment, but Chōzō is unrelated to them. We can imagine that the boy can open his heart to Chōzō, although he had closed it to his parents.

What I have said indicates the deep implications encounters can have in our lives. Chōzō is probably not worth much as a person, and as a member of society he is of no consequence whatsoever. But in terms of this boy, encounter with a total stranger becomes one of the most meaningful experiences of his life. We could say it is almost therapeutic because it causes him to re-examine himself and heretofore unperceived implications of his own behavior. He recognizes, for instance, the inconsistency inherent in the fact that all the time he is feeling emotionally dependent upon Chōzō he insists he can find his own way to the mines. He adds, "I have developed a theory since this happens to me often. Just as every illness has an incubation period, there is an incubation period to our thoughts and feelings. During this time we are unconscious of the way in which they possess and dominate our minds and, so long as outside events do not bring them into consciousness, we go through life denying their influence. But the proof of their existence lies in the very efforts we make to deny them, and seen objectively these efforts appear contradictory. Sometimes they even strike the person himself that way. At any rate, they can entail a great deal of suffering. All of my suffering over the girl I mentioned before . . . at the heart of it was my inability to realize what was incubating inside of me."

It is extremely interesting to note that the boy's theory closely resembles the psychoanalytic concept of repression and that it enables him to look at the events that led to his leaving home in an entirely different light. In the past others were to blame for his unhappiness. Now he is able to see the problem as one arising within himself and as one that could have been prevented.

Of course, he only dimly perceives this. With more time and greater insight he would have returned home. In the meantime he can only do as Chōzō asks. He describes his frame of mind this way: "My spirit was listless, like a body hung over from drink. It was drugged with sleep." Thus, when he steps off the train and stares at the scene before him, he experiences "a very peculiar feeling. While my mind perceived the brightness of the world outside, it did not function well enough to perceive it as real. The sharp, straight lines of the road and the buildings appeared as a dream that closely approximated the real world." These are feelings of depersonalization. The boy sees the reality before him and knows that it is there to be dealt with. "But I had lost the will to do it. I stood there with only the vivid visual impressions registering on my eyes." There is hardly a psychiatric textbook that gives a more astute description of feelings of depersonalization.

Eventually the two arrive at the mines. Chōzō introduces the boy to the head of the mess hall and vanishes. The boy is overcome with feelings of helplessness, but it is too late to return home and, while he is determined to become a miner, he must endure foul gruel, bedbugs, and the derision of the workers. Moreover, the following day he descends into the mine and is terror-stricken at the realization that death, heretofore remote, is close at hand. He experiences a faint ecstasy when his guide leaves him momentarily in a dark tunnel, but in the next instant he realizes that "This is death. How awful to die here!" and the spell is broken. Again, as he is climbing out of a lower shaft, he is tempted to release his grip on the ladder, but a voice tells him, "This is a dreadful place to die. Get out of here and jump over Kegon Falls."

He becomes separated from his guide and, while searching for an exit, he encounters a miner named Yasu. Yasu inquires about the boy's background and, after telling his own life story, strongly urges him to return to the everyday world. He even promises to raise the money for the trip. While the boy recognizes that, unlike the other miners, Yasu is a highly likable man of education and nobility, he does not feel he can accept his offer. "When I recall it now, my refusal stemmed from a feeling that, considering who he was, it would be shameful . . . it would be degrading to accept his money. He was an extremely fine person, and I was determined to be as fine. Not to be would have meant losing face. How nice it would have been to accept his offer only to please him. But how like a beggar to take what was not mine for a purely private end. I stood a pauper before one as worthy of respect as Yasu. I could not bear to give him proof of it." The boy feels undeserving of the money and, so great is his respect for the man and his desire to be like him, he cannot leave. When he hears Yasu's own tragic story, moreover, he experiences a profound pity for the man. We can say that in one way or another the boy comes to identify with him. As proof, we need only look to a later passage in which he says that if Yasu goes on living and working so must he.

But what about Yasu? Why does he demonstrate interest in a youth he has never seen and attempt, by defraying his traveling expenses, to send him back to society? Yasu admits to having been inextricably involved in the commission of a crime but, because he cannot believe that he had committed a wrong act, he is too proud to entrust the matter to the courts and even to this day remains in hiding. Although he no longer wishes to return to society, he feels the unfathomable pain of a person dead to the world. Who should suddenly appear before his eyes but a youth gone astray. Yasu persuades him, "Go home while you can. The road ahead runs downhill for me . . . but that would be terrible for you." We can interpret these remarks as Yasu's desire to restore the youth to society as his proxy. Just as the youth identifies with the man and wishes to be his equal, the man likewise identifies with the youth and finds in his salvation a new reason for living.

As I stated previously, the boy does not heed this advice. He renews his resolve to stay and even seeks Yasu's consent. Actually, he decides to return home after five months, and it is probably Yasu's direct and indirect influence that persuades him to do so. Thus, the encounter with Yasu becomes the decisive turning point in his life. I said before that the encounter with Chōzō was almost therapeutic. The encounter with Yasu is therapeutic in the truest sense of the word. While Chōzō becomes a temporary object for the boy's dependency need, Yasu becomes, as an index of character, a permanent object for identification.

Our age abounds with men and women who, like this boy, attempt to flee life completely. If these people are to go on living, they must find a permanent object for identification. If that object is to exercise its fullest effectiveness, it must like Yasu be able in some degree to identify with them. We can even think of this as a fundamental principle for all forms of guidance work.

Chapter Three

SANSHIRŌ

1908

Synopsis

At age twenty-three, Ogawa Sanshirō has graduated from Kumamoto Higher School in Kyushu and is leaving his widowed mother to go to Tokyo to enroll in the Imperial University.

His troubles commence on the train. A young housewife asks for help finding accomodations at the overnight stop in Nagoya. She maneuvers him into sharing not only a room but a bed and, when the night passes uneventfully, laughs at him for "lacking pluck." Back on the train he encounters a strange looking man who, in a droll critique of the times, announces that Japan is headed for ruination. No one had ever dared be so outspoken in Kumamoto, and Sanshirō is embarrassed by his own naive opinions.

Tokyo is a fearsome place and, when Sanshirō becomes homesick, he visits Nonomiya Sōhachi, a friend from Kumamoto who works in a university laboratory. The scale of Nonomiya's research on the "pressure of light rays" proves overwhelming and, hurrying away, Sanshirō seeks the solace of a pond situated in the middle of the campus. There he sees a beautiful girl silhouetted against the trees on a rise above the pond. They exchange glances when she passes near him, but he is inhibited from speaking to her by the memory of the woman on the train.

School begins and Sanshirō attends faithfully. He meets a classmate named Sasaki Yojirō, and Yojirō introduces his country friend to the wonders of the city. Sanshirō also learns that Yojirō lives with the strange man he met on the train. His name is Mr. Hirota and he teaches English literature at Tokyo Higher School.

One Sunday when Sanshirō visits Nonomiya's house in the suburbs, a telegram arrives from Yoshiko, Nonomiya's sister, who is in the hospital. Sanshirō agrees to watch the house while

Nonomiya goes to see her. During the night a young woman commits suicide on the railroad tracks that run near the house. Sanshirō begins to fantasize that the suicide victim is somehow romantically involved with Nonomiya and that Nonomiya's sister is none other than the girl by the pond. He hurries to the hospital the next morning to meet Yoshiko, only to be disappointed. Turning to leave her sickroom, whom should he see framed in the light of the hospital entryway but the mysterious girl. She asks for directions to Yoshiko's room. He notices a ribbon in her hair he once saw Nonomiya purchase.

The mysterious girl appears again when Sanshirō is asked to help Yojirō and Mr. Hirota move to new lodgings. She introduces herself as Satomi Mineko; her older brother is Hirota's former student and a friend of Nonomiya. The two youngsters enjoy working together and, as they arrange Hirota's library, Sanshirō mentions the terrible condition of the university library collection. Not even the works of Aphra Behn have escaped pencil markings, he says. Hirota is impressed that Sanshirō knows this minor seventeenth-century British novelist and, during lunch, he challenges his young helpers to translate a line adapted from Behn's Oroonoko. *The phrase—"Pity's akin to love"—seems Japanese in feeling but defies translation. To the amusement of all but Hirota, Yojirō attempts a colloquial rendition: "To say 'You poor thing' is to say 'I love you.'"*

Shortly thereafter, Mr. Hirota and his young friends take an outing to the chrysanthemum displays held at Dangozaka. When Mineko is overwhelmed by the crowd, Sanshirō leads her to a nearby brook. This is their first chance to be alone together, but Sanshirō soon grows nervous about rejoining the others. Mineko reminds him that, while they may be "lost children," they are big enough to fend for themselves. She adds cryptically that the English term for "lost child" is "stray sheep." This leaves Sanshirō all the more baffled.

Yojirō does not join the group that day but stays home working on a manuscript. He has decided that Hirota deserves more recognition than he receives as a higher school teacher and, borrowing twenty yen from Sanshirō, he publishes an article calling for the ouster of the university's Professor of English and replacement with "The Great Obscurity," the name he coins to identify Hirota. When he defaults on the loan and Sanshirō is unable to pay his rent, Mineko agrees to lend Sanshirō the money if he will come to collect it in person.

The money is of little importance to Sanshirō when he calls at her house. What he intends to say is that he has come first and foremost to see her, but he is unable to summon the courage. After a trip to the bank, he and Mineko attend an art exhibit where they chance upon Nonomiya. When Nonomiya suggests with a snicker that Mineko and Sanshirō are an unlikely couple, Mineko retorts that they make a handsome couple.

Her retort does little to allay Sanshirō's anxiety about his relationship with her. He has seen her recently in Nonomiya's company at a school athletic meet, and Hirota and Yojirō have warned him that Mineko is like an Ibsen heroine who will marry only a man she admires. He also hears, however, that when Mineko was asked by Hirota's friend Haraguchi, an artist, to sit for her portrait, she insisted on posing in the outfit she wore the day by the pond.

When money arrives from home, Sanshirō calls at Haraguchi's studio. Mineko's pose brings back to him the full force of their first meeting and, when the session is over, he succeeds in saying what he could not say before: "I came because I wanted to see you." Mineko is so exhausted from the sitting she hardly pays attention; and then a man in gold-rimmed glasses appears in a rickshaw to take her to another engagement.

Yojirō's scheme has in the meantime backfired. Hirota is denounced in the newspapers for using his students to promote his career, and Sanshirō is mistakenly identified as the author of the offensive article. Sanshirō calls at Hirota's house, and teacher and student commiserate together.

Several days later Sanshirō learns that Mineko is engaged to the man with the gold-rimmed glasses. Waiting outside the church she attends, he asks her plans. She says softly, "I acknowledge my transgressions and my sin is ever before me."

The New Year holiday passes, and Sanshirō returns to Tokyo after a visit home. Mineko has married, and her portrait, "Woman Standing in a Wood," hangs on exhibit. Hirota, Nonomiya, Yojirō, and Sanshirō go to see it. When Sanshirō objects to the title and is asked to suggest a better one, the words "stray sheep, stray sheep" catch in his throat.

Sanshirō's Task

Sanshirō has graduated from Kumamoto Higher School and is in high spirits as he sets forth on his journey to Tokyo to enroll in the humanities department of Tokyo Imperial University. He would not have believed that on the way he would share a bed in a Nagoya inn with a strange woman or, to make matters worse, that she would laugh at him for failing to turn the situation to his advantage. Nor would he have imagined that on the connecting train an eccentric looking man would confound him with his startling remarks or that, upon arrival in Tokyo, the uproar of the city would prove totally overwhelming. "Everything looked as though it were being destroyed and rebuilt in the same moment." Tokyo struck Sanshirō in exactly the same way it strikes us half a century later.

It is, of course, his non-involvement in the furor—rather than the furor itself—that Sanshirō finds truly alarming. He says, "The world is in a state of constant upheaval. I see the upheaval but I cannot join it. My world and the real world lie side by side, but nowhere do they intersect. It is very distressing to know that as the world advances I will be left behind."

In the midst of this uneasiness, a letter arrives from home. It strikes him "as having arrived from the musty past," but when he considers that his mother is his sole link with the real world he follows her advice and calls on Nonomiya Sōhachi, a hometown friend studying in the university's science department. There, he experiences a fleeting sense of admiration for the quiet, detached world of the scholar. At the same time, he is haunted by the utter loneliness of being a man about to launch into life. Hurriedly excusing himself, he wanders about the campus, where, by the pond, he encounters a beautiful young girl. Their eyes meet for a moment, but the memory of how the woman on the train had laughed at him suddenly returns and he is afraid. He confusedly mutters the word "contradiction."

In the days that follow Sanshirō busies himself attending lectures as he believes every good student should. Still, something is missing from his life. He meets Yojirō, a classmate; and his new friend introduces him to the streetcar system and the vaudeville theaters. Yojirō even teaches him to use the university library. One day, in the midst of all this activity, Sanshirō is asked to deliver a package to Nonomiya's sister, who is in the hospital. As he is leaving her sickroom, the young girl he had seen by the pond comes to call. Thereafter, his mind is captivated with the thought of her and he cannot settle down.

He hears that Yojirō is boarding at the residence of a certain Mr. Hirota, Nonomiya's teacher in higher school. The more Sanshirō learns about Hirota, the more he imagines him to be the same man he had met on the train; and, surely enough, he is. As time passes Sanshirō comes to be included in the group of young people who have gathered around this man Yojirō calls "The Great Obscurity."

One evening, after reading a second letter from his mother, Sanshirō gets into bed. As he lies there thinking, he realizes that his surroundings are constructed of three worlds. First, the world of his hometown, inhabited by his mother and Omitsu, the girl whom she hopes he will marry. This world "was far away. It had the redolence of what Yojirō called 'Old Japan.' All was tranquil, but all was lost in sleep. To go back, of course, required no effort on his part. He need only go. It was a sort of refuge, and he would never feel the need of it except in the most extreme circumstances." The second world is the world of academic endeavor, and it is dominated by Hirota and Nonomiya. The third is the colorful domain of the opposite sex which seems to entice and reject him at the same time. Having fantasized at some length about these three worlds, Sanshirō "put them in a row and compared them. Next, he mixed them together and abstracted a single result. Nothing could be better than to bring his mother from the country, marry a beautiful girl, and devote his life to academic pursuits." Relieved for the moment, he falls asleep.

So far I have summarized the novel up to the point where the major characters have made their appearance. We now have a rough idea of Sanshirō's mental and physical circumstances. He has left his family to come to Tokyo. He is a freshman in college; his future looks bright, and his aspirations are many. Yet his initial experience is one of disillusionment. There is a great gap between himself and the real world. The city is like a great machine that is relentless in its motion and indifferent to his fate. Both the type of study and the type of woman he admires lie far beyond his reach. He knows he needs the real world, but it seems "dangerous and unapproachable." He has taken the ambitious step of leaving home only to discover to his frustration that his is a restless soul that does not know what it wants.

Actually, Sanshirō's state of mind is common to all adolescents. As the caterpillar must pass through the amorphous pupa stage in order to become a butterfly, a child too must experience the unsettled period called adolescence to become an adult. It happens that Sanshirō finds himself right in the middle of adolescence. Indeed, his sole contact with reality may be through his ties with home, but he is no longer home. He may miss his mother, but he is no longer a child who will do as she says. But neither is he an adult. The world he aspires to is far away. He lives in the city, but he has yet to become a man of the city.

It is highly significant that in this set of circumstances he refuses to reject his past and continues to recognize its value as a refuge. It is also significant that he does not absorb himself in a mad attempt to force society to recognize his existence. Few adolescents today have such emotional flexibility, and perhaps they might envy Sanshirō who can retreat into himself and devise a unified vision of his three worlds. Admittedly it is, as he says, extremely commonplace, but I do not believe that we can laugh at Sanshirō's naïveté. I would prefer to esteem the sincere expenditure of effort—"He had given the matter considerable thought before arriving at this conclusion"—that went into creating it.

Psychoanalysis has a concept known as ego identity which American psychoanalyst Erik Erikson defines as follows. Ego identity "...express[es] the essence of the ego gain which a youth must accomplish: namely, a need to create a continuity and sameness out of, one, what he was as a child and is becoming in the present and, two, what he conceives himself to be and what the community sees in him and expects of him."[9] Adolescents who do not achieve this continuity go emotionally bankrupt. In this respect all youth are the same whether they be the Sanshirōs of the Meiji era or contemporary youth in Japan and the West. Sanshirō is about to confront the central task of adolescence.

"Pity's Akin to Love"

Helping Mr. Hirota to move provides an opportunity for Sanshirō to meet and become friends with Mineko, the girl whose image has been fixed in his mind ever since their chance meeting by the pond. One afternoon Hirota and his young friends go to the chrysanthemum displays. When Mineko becomes ill and leaves the group, Sanshirō follows her. This is the first opportunity the two have to be alone, and during their conversation Sanshirō's attention is drawn to the words "lost children,"—and Mineko's English translation, "stray sheep,"—which she repeats in a suggestive manner. Her meaning eludes him until several days later when a postcard arrives. Two lambs are depicted lying by a brook; standing over them is a sinister looking man labeled "The Devil." The card is signed simply "Lost Child." Sanshirō immediately recognizes that it is from Mineko, and he is delighted at her metaphor comparing them to two lost sheep. "At last he understood the meaning of 'stray sheep.'" In other words, he believes she is indicating partiality for him.

He has reservations, nonetheless. He wonders if she "is not too formidable a match," and he cannot escape "the twinge of humiliation that comes from being seen through"—a feeling that grows stronger when he realizes that she has already detected his anxiety about her relationship with Nonomiya. After all, he tells himself, he cannot compete with Nonomiya. "His mind was still unschooled and his views unformed." The strain of wondering if Mineko is toying with his feelings eventually becomes almost intolerable.

Chance provides an appropriate opportunity to dispel Sanshirō's doubts. Mineko agrees to cover a debt that Yojirō has incurred with Sanshirō, and she stipulates that Sanshirō is to come in person for the money. He looks upon the offer as evidence of her partiality, although it does not wipe away his fears completely. On the day that he calls at her house, their conversation proceeds very pleasantly. He is charmed to the point of abashment by her comment that "[he is] an easygoing person who would not question another's feelings even if they came replete with index" and by the special air of intimacy she displays toward him that afternoon when they happen upon Nonomiya at the art exhibit. He does not feel more confident of having ascertained her true feelings for him, however.

Later, the loan provides a second opportunity to delve into her feelings; and, on the pretext of making repayment, Sanshirō visits Mineko while she is sitting for her portrait. When the session is over and the two leave the studio together, he states his real reason in coming: "I came to see you." Although he repeats this admission twice, it fails to elicit a favorable response; and, to make matters worse, a distinguished looking young man appears in a rickshaw and whisks Mineko away to a prior appointment. It is not long after that Sanshirō learns from Yojirō that Mineko is engaged.

To date, a great deal of criticism has been written interpreting Sanshirō's love affair and heartbreak as a result of either Mineko's "unconscious hypocrisy"[10] or his own vanity and naïveté. I would like to try, however, discussing it in different terms. That is, in terms of the phrase "Pity's akin to love" that appears in English in the Japanese text. Hirota quotes this line from a play by an eighteenth-century English dramatist[11] on the day when Sanshirō and Mineko first meet at his new lodgings. I cannot help feeling that Sōseki deliberately inserted this phrase as a key to understanding their relationship. The novel says that when Yojirō translates it into a brand of Japanese used in popular ditties of the period (i.e., "To say 'You poor thing!' is to say 'I love you'"), Hirota lambastes the translation for its vulgarity. The important point is that Yojirō glosses over the meaning of "akin," making pity equivalent to love. We can interpret Hirota's objections as pointing to Yojirō's failure to make this distinction.

Whether it be Sanshirō's feelings for Mineko or vice versa, both represent fairly complex sets of emotions. I think there is little question that at least Mineko's derive chiefly from feelings of compassion and understanding. Her comparison of their relationship to two lost sheep is the frankest indication of this fact. She knows only too well the confusion Sanshirō is experiencing about life and that his confusion even extends to his feelings about her. I do not doubt that Sanshirō is greatly attracted to Mineko, but he clearly vacillates between thinking of her as a temptress who toys with him or a goddess come to save him. Mineko must be either the answer to all of his problems and worries or, if not, someone on the order of the woman on the train.

We can imagine that Mineko has her confused moments too. The novel does not portray them, but her ability to understand and sympathize with Sanshirō more than likely reflects the presence of doubts in her mind. It is altogether possible that she is also attracted to Sanshirō, but, as her postcard indicates, she considers their relationship to be more like two lambs than two lovers. Truly, sympathy is akin to love but is not the same. Can it not be said then that, having recognized the true nature of their relationship, she secretly hopes Sanshirō will understand her in the same light that she understands him? This is indeed the message she tries to convey when, having raised the subject of stray sheep to Sanshirō's baffled silence, she has to ask, "Do I seem too smart to you?"

Often to our surprise Westerners say, "I like you but I do not love you." We think that to like a person is to love him, whereas Westerners make a clear-cut distinction between these two feelings. Just as the word "like" is used in the English language in the dual senses of "to be fond of" and "to be similar to," liking indicates both a closeness and an affinity. Loving, on the other hand, is to love an object different from oneself. I do not wish to belabor this point, but because in general Japanese tend to think that "to like" is "to love" and because we agree with Yojirō that to say "You poor thing!" is to say "I love you," our sympathy goes out to broken-hearted Sanshirō and we want to condemn Mineko, the "unconscious hypocrite," for having apparently cast him aside. That is why a reading of this novel never fails to induce a feeling of glorious intoxication: we imbibe once again the joys and sorrows of our lost youth.

Those Who Are Left Behind

I have already described the anxiety Sanshirō experiences at being left behind by the ever-changing world. As it happens, fear of abandonment reappears in a number of different situations in the novel, and it might be considered a central theme. For example, when Sanshirō meets Nonomiya for the first time, Nonomiya says, "Research progresses at such a pace these days that, if one is less than careful, *he'll soon be left behind.* People may think that I do nothing in this cellar laboratory, but the inside of my mind is feverishly at work. Who knows? It may be working harder than one of those new streetcars. I don't even like to take time for a trip in the summer." [Doi's italics] Yojirō frames the same idea in a different way. "Can thoughtful men standing at the center of intellectual circles afford to be indifferent to the violent upheaval taking place in people's thinking today? Since the power of letters is in the hands of youth, wouldn't we be committing a serious omission not to make the most of it? The literary world is undergoing a precipitous and startling revolution. Everything, everything without exception, is heading in a new direction, and *there will be hell to pay if we get left behind.* We must seize the initiative for ourselves." [Doi's italics]

While joining the Hirota group does not eliminate Sanshirō's anxiety, it does contribute greatly toward making his life more complete. This is best illustrated by a passage describing his day at the chrysanthemum displays. "Sanshirō felt that his life was more meaningful than it had ever been in Kumamoto. Of his three worlds, the second and third were represented by the personalities of this group. Half of the members were a dull, scholarly gray; the other half, bright as a field of flowers—a happy and harmonious combination, he thought. Moreover, in no time he too had been woven into this tapestry. Still, something was not quite right, and he worried about it."

What troubles Sanshirō is his suspicion that he has been excluded from a special relationship internal to the group; or, more specifically, he is jealous of Nonomiya's relationship with Mineko. Toward the end of the novel his suspicions narrow and focus solely on Mineko, who does finally alienate him. She literally leaves him behind, and the novel closes with him seated in front of her portrait in the exhibition hall saying to himself, "Stray sheep, stray sheep."

It is perhaps the irony of this novel that the character who dreads abandonment most is actually abandoned. Interestingly enough, however, we also find Sanshirō leaving others behind, namely, his own family. His first letter from home seems to have "arrived from the musty past"; and, in the passage in which he describes his first world as a refuge, he says that it does not seem fair that his mother should be buried away in such a place. Of course, in going to the university he does not deliberately mean to leave his mother behind, but, as his own words indicate only too well, he is, psychologically speaking, deserting her.

It is also interesting that, as his relationship with Mineko becomes more troubled, his longing for home—his refuge—grows. His mother's handwriting is a welcome sight; and in one reminiscence he mentions that he sleeps better at home. His sudden nostalgia is one instance of regressive behavior. From the time Sanshirō arrives in Tokyo he is apprehensive about being left behind by the world or being jilted—ample evidence that he is in a mental state ready to engage in regression.

We can say this because the fear of being left behind is fundamentally an infantile feeling. A child's greatest joy is being with the mother; its greatest fear is being abandoned. It behaves in such a way as to keep the mother nearby and it retains this pattern of behavior into adult life. We might say, for example, that the tendency of many Japanese to be overly anxious about the need to be considered fashionable is one manifestation of this pattern. I do not wish to diminish the importance this desire plays in helping people to mature, but the individual who forever feels the need to keep up with others is in serious trouble. For when his efforts prove ineffectual he easily falls prey to anxiety against which he has no defenses. More than likely, Sanshirō is this type of person. Initially he fears being left behind and, in spite of his efforts to the contrary, he finally is.

Thus, outside of his mother's letters, his only refuge is Mr. Hirota. This man makes him feel warm and comfortable; in his company Sanshirō is "carefree; the need to compete seemed less important." The view that Hirota espouses on the train—"Tokyo is bigger than Kumamoto. Japan is bigger than Tokyo. And the inside of your mind is bigger than all of Japan"—shocks him, but it also helps him forget his anxiety for a short time. It could not, unfortunately, get to the root of his problem.

Note in contrast that another character—another stray sheep—maintains a different relationship with Hirota. Certainly Mineko is quite close to him, but there is nothing to suggest that she is startled by his statements or that she ardently admires him. Rather we find it is Hirota who is exclaiming that "that girl is perfectly calm . . . and violent!" As a matter of fact, it appears that through marriage Mineko leaves behind not only Sanshirō but also Hirota and Nonomiya. What is Sōseki telling us when he has her quote an Old Testament Psalm as her parting words to Sanshirō? "For I acknowledge my transgressions, and my sin is ever before me" are hardly the words of an "Ibsen girl" as Yojirō had once described her.

Chapter Four

SOREKARA (And Then)

1909

Synopsis

A man of taste and intellect, Nagai Daisuke lives with a maid and houseboy in a comfortable residence in Ushigome, Tokyo, and, at age thirty, leads the life of what Sōseki calls "an educated idler."[12] *He does not work because his father has provided for him ever since his graduation from the university.*

Daisuke believes there are few opportunities for meaningful work in a day and age when society has become obsessed with the pursuit of individual and national aggrandizement. He feels that his time is better spent thinking and reading. His health too is a matter of concern and he prefers to avoid any stimuli that may tax his delicate nerves. The color red, for example, always makes him wince.

Daisuke's father, Nagai Toku, espouses a more vigorous life based on old-fashioned Confucian precepts of hard work and devotion to family and country. As a young samurai he participated in the Meiji Restoration and, in the decade since retiring from government service, he has established a prosperous business. He is often angry with his son, and Daisuke is, in turn, distressed by the extent to which material success appears to have corrupted his father.

As the novel begins, Daisuke's friend Hiraoka Tsunejirō has returned to Tokyo. The two men were close friends in college, and three years earlier Daisuke arranged for Hiraoka to marry Suganuma Michiyo, the sister of a mutual acquaintance. Daisuke also loved Michiyo, but he relinquished her to Hiraoka in a "chivalrous gesture." The newlyweds moved to Kyoto in connection with Hiraoka's work as a banker. They were happy at first, but then Michiyo lost her baby and now Hiraoka has been forced to resign from his job and make good the loss of a thousand yen supposedly embezzled by a subordinate. Daisuke puts his houseboy to work finding the couple accommodations, and he approaches Seigo, his older brother, about a loan and a job for Hiraoka.

When Seigo refuses, he turns to his sister-in-law. Umeko is the only member of the Nagai family who appreciates Daisuke for his intellect, and she often intervenes to protect him from his father's anger. This time she refuses to help; or at least until the following day when she sends a draft for part of the loan. She also presses for a reply to his father's latest choice of a bride for his son. The girl is well-to-do and related to a man who once interceded to save Mr. Nagai's life. As usual, Daisuke is not interested. Asked whom he would marry, the name Michiyo flits through his mind.

In spite of Daisuke's efforts, Hiraoka is not at all grateful when he calls, sporting a new suit obviously purchased with the loan. He says perfunctorily that Michiyo will come later to express her *thanks. Daisuke dislikes the way his friend has changed, but what distresses him most is Hiraoka's indifference to Michiyo and the state of her health. She is suffering from a serious heart condition. Several days later she arrives breathless at his door, having run to avoid a sudden rainstorm, and collapses into an armchair in his study. She brings a bouquet of lilies, his favorite flower. He notices that her hair is dressed as it was the day they first met years before.*

Meanwhile, Mr. Nagai has pushed ahead with negotiations for his son's marriage. When Daisuke is called out for a first, and then a second surprise interview with the prospective bride, he decides to take a trip to avoid having to meet her again. He visits Michiyo before leaving Tokyo. She is in bed, and her health so poor that he gives her his traveling money and decides to speak to Hiraoka. Hiraoka has found work reporting for the financial section of a newspaper, but he stays out nights and spends his salary on himself. Imagining that Daisuke has come to collect the loan, he ignores his entreaties about Michiyo's health and insinuates that he is suppressing a scandal concerning the Nagai family enterprises.

Forced to decide whether to take Michiyo from Hiraoka or

to marry the other girl, Daisuke calls Michiyo to his house and confesses his love. She weeps and demands to know how he could have originally forsaken her. He is sorry but says that she has had her revenge by making it impossible for him to love any other woman. After Michiyo leaves, he takes the lilies arranged in his study and, stepping into the moonlight, scatters the petals in the garden. "It is finished," he says to himself.

He grows more and more anxious as the day approaches to reply to his father. Mr. Nagai accepts his son's refusal quietly; nothing is said of Michiyo. Next, Daisuke confronts Hiraoka who, ironically enough, has been at home caring for Michiyo who has collapsed from nervous exhaustion. Hiraoka agrees to surrender her but, greatly aggrieved at his friend's betrayal, stipulates one condition: Daisuke is not to see Michiyo until she recovers. Daisuke leaps from his chair and wildly accuses Hiraoka of planning to deliver a corpse. Convinced that Michiyo is dying, he makes several furtive trips to her house. It is always dark and empty.

Several days later Seigo calls on Daisuke. He has a letter from Hiraoka. Daisuke admits to the veracity of its accusations and Seigo, dumbfounded at his brother's behavior, bids a last farewell and pronounces him forever disinherited.

Daisuke lingers a moment in his study before announcing that he is stepping out to find a job. The hot June sun beats down on him as he boards the streetcar, and his head begins to spin. The bright red color of a mailbox catches his eye and, in an instant, every red object in the surroundings leaps out as if to attack him. Daisuke's world seems about to be enveloped in crimson flames.

Daisuke's Self-Image

Daisuke has just awakened as the novel opens. He is holding his hand over his heart to check its beat. This is "his habit of late," and the little ritual assures him he is very much alive. He trembles at the thought that life is separated from death by only a heartbeat. Next, he climbs out of bed and steps into the bathroom. There is a brief and approving inspection of the even alignment of his teeth and the healthy glow of his complexion; for a moment his eyes fix fondly on the image in the mirror as he examines the part of his hair, the cut of his moustache, and the fullness of his face. "The patent fact of being alive struck him on occasion as a miraculous stroke of good fortune." He takes pride in his appearance and does not hesitate to use a little powder if necessary. "It did not bother him in the least to be called a dandy." Daisuke has "outgrown Old Japan."

A psychiatrist might have a word to say about the state of mind described in the opening passage of this novel. He would concur with an opinion expressed later the same morning by Kadono, Daisuke's houseboy, who says, "If you are too concerned about your health, sir, you will take ill for sure." Although Daisuke makes light of this view—"I am already sick"—his undue concern with his health is symptomatic of hypochondria; and his fascination with the face in the mirror, of narcissism. An experienced analyst would also note that, like the case of Narcissus in Greek mythology, Daisuke's preoccupation with himself probably dates from the time he was alienated from a person he should have been close to.

Of course, Daisuke is not able to analyze his thinking in this light. He seems partially aware of an abnormal nervousness, which he attributes to "the penalty one pays for being endowed with a fine mind and keen senses. It is one of the disadvantages attendant on a good education." He believes moreover that learning to live with this handicap has made him a better man; at times he is convinced this is the point of his life. In short, Daisuke places enormous confidence in his own judgment. The fact that he has outgrown Old Japan is a source of pride, not anxiety; and he considers that, despite certain drawbacks, his self-image is his most valuable possession. We can say that the novel asks two questions about this image. What does it derive from? Will it stand the test of a lifetime?

The second son of a successful businessman, Daisuke is thirty years old. While his older brother holds an important position in their father's corporation, he himself does not work. We know that his mother is dead, although there is no indication of when she died. As Daisuke is the youngest, it may be that she died when he was small. He is extremely fond of his sister-in-law, Umeko, who acts as a kind of mother surrogate. He has recently moved out of his father's home and set up a separate residence with a live-in maid and a houseboy. As he has no income, however, he calls at his father's at least once a month to receive an allowance; sometimes he visits to play with his niece and nephew or to chat with his sister-in-law. In other words, Daisuke is still living off his father at an age when he should be financially independent. It seems that this sort of carefree existence was still possible as late as the Meiji era, although I should not like to suggest that it escaped censure. Actually, Daisuke's father frequently lectures his son on his duty as a Japanese subject to engage in purposeful activity. Daisuke pays little attention because in the last few years he has grown increasingly skeptical about the integrity of his father's character. As for the integrity of being a financial dependent, he never raises the question; and his father remains unstinting with his funds despite his annoyance.

Daisuke views his financial well-being as proof that he belongs to a "superior class of people possessed of hours unspoiled by work." Although he has no job, he flatters himself that he has led a constructive life for the last few years. Prior to that time he had been proud of the Confucian work ethic he had inherited from his father and made a self-righteous display of it. In the bottom of his heart, however, it seemed unreal. He was merely "the gilding that had rubbed off of his father . . . It was distressing being gilt when everyone—his father, his seniors, anyone with education—gleamed like gold. He was in a hurry to be gold too." As his "unique powers of observation and thought began to peel away his own gold plate," his father, as well as the others, suddenly lost his glitter. Daisuke's skepticism about his father dates from this period.

In psychoanalytic terms we would refer to this change as casting off the parental superego. In adolescence people often reexamine the morals they learn from their parents and revise them in favor of a more realistic code. Daisuke differs, however, in that he rejects his father's morals *in toto*. It also happens, tragically enough, that his father's moral code coincides with that of society at large, and by calling one into question he must reject both. He is left, so to speak, in a moral vacuum—a situation that represents a serious emotional crisis. One imagines that this crisis is the source of his neurotic tendencies toward narcissism and hypochondria.

How conscious he is of the severity of his predicament is open to question. As we saw earlier, he expects to suffer on account of his fine mental faculties, and he is confident of his ability to provide intellectual solutions. It is this conceit, in fact, that prompts him to turn to his friend Hiraoka, whom he has not seen for three years, and say, "I cannot think of anything more worthless than what people call the lessons of experience. They entail hardship and nothing else . . . Struggling for one's daily bread may be terribly important but it is inferior work. A man has not lived until he has the luxury of a life removed from the need to provide for food and drink. You may think me a child, but when it comes to living with affluence I am considerably your senior."

One wonders why Daisuke feels the need to brag and talk competitively. It takes a very conceited, or a very cruel, man to talk this way to a friend who has lost his job and returned in defeat. Daisuke does not, however, realize the implications of his own words. For surely his conceit stems from a desire to show Hiraoka how he has come to understand the basic problems of life rather than from any desire to belittle him. Just as Hiraoka has changed as a result of his hardships during the last three years, Daisuke too has changed inside; and the change has been dramatic enough to produce neurotic tendencies. Naturally, he is too proud to admit to having emotional problems. He must pretend a great deal, and it is this pretense that causes him to brag. He is, nonetheless, sick inside and, although he is still able to relate to his family and friends, inwardly he is showing signs of alienation. Thus, the hypochondria of late.

"It's the Fault of the World"

Later Hiraoka has his small revenge. "Yes," he says, "I admit to failure, but I was working and will keep on working. You look at me and laugh. And even if you say you don't, am I not right in thinking you might as well be? You laugh, but what are you doing? You take the world as it comes. You're the type who refuses to try to change it. But the urge still exists because, after all, you are a human being; and your refusal to gratify that urge explains why you always feel empty inside. I wouldn't be able to live without the feeling that, as a result of asserting my will, the world is even to a limited extent the way I want it. That is what I see as the purpose for my life. But you only think, with the result that you construct two separate worlds—the one inside of your head and the one outside—each with a life of its own. Hasn't it ever occurred to you that living with that contradiction is the greatest failure of all? Go ahead, ask yourself why. At least I push the contradiction outside of myself, and that probably makes my failure a little less serious. You laugh at me, but I can't laugh at you. Much as I would like to, society shakes its finger at me and says don't you dare."

This is scathing criticism. Hiraoka is sharply critical of the fact that Daisuke has isolated himself from reality through an excessive reliance on his intellectual powers; or, to state that in psychoanalytic terms, that he has used intellectualization as a defense mechanism in order to withdraw into a world of ideas and convince himself that he has solved all of life's problems. Daisuke will not submit to this view, however. It would mean the destruction of the world he has struggled to create, as well as a serious loss of confidence in himself. Consequently, he refuses to retract his earlier statement and consider the wisdom of finding a job. He explains why when the subject is raised a second time. "That I do not work is not my fault. It's the fault of the world. To put it in larger terms, it is the whole crazy relationship between Japan and the West. No other country looks more like a pauper and has as many debts as we do. And when do you think these bills will be paid? Yes, we can pay off the foreign bonds, but that is not all we owe. We're a nation that cannot survive without borrowing, but here we are talking like a first-class power and trying to do the impossible of joining those that are. No matter where you turn, we have skimped on substance in order to throw up a façade of power. And what makes it sadder still is the half-baked way in which the job has been done. We're like the proverbial frog who tried to blow himself up into a cow, and one of these days we are going to burst. Don't you see what is happening to all of us as individuals? People do not have time to relax and think and accomplish something worthwhile. What with the content of education slashed to the bare minimum and the way people are overworked, we will end up with a nervous breakdown. Talk to people, and you will find they are fools. They think only as far as themselves and their own needs. And who can blame them since they are too tired to think? Spiritual fatigue and physical atrophy go hand in hand—along with moral decline. Search throughout Japan and you will not find a square inch that inspires optimism. Everywhere gloom. Standing in the midst of such gloom, there is nothing I can say or do that will make a difference.

"All right, I am lazy, but I have been lazy ever since you've known me. I used to force myself to be active, so perhaps you thought I had a glorious future ahead of me. I still would if society were for the most part physically, morally, and spiritually sound. There would be any number of things I could do, and there would be stimuli enough to shake me from my inertia. But, the way the world is now, I would rather be by myself. Taking it as it comes, as you say, and deriving my satisfaction from what suits me. I certainly will never convince anyone to think as I do, however."

Daisuke's critique of modern Japanese society is no less penetrating than Hiraoka's analysis of Daisuke the individual. In comparison with Japan in his day, contemporary Japan is far wealthier despite, or rather because of, its defeat in the last war. Yet this change of affairs may represent mere window dressing and, as Daisuke would say, an increase in debts to the West. A fundamental solution in respect to Japan's relations with the West is still to be achieved, and that is why it is safe to say that statements of contemporary social critics do not differ appreciably from what Daisuke has already said about the distortions resulting from westernization and modernization. In terms of social criticism, then, his argument posits a number of fine points. We must not forget, however, that it is a personal apologia. For when he turns to Hiraoka's wife for confirmation—"Michiyo, what do you think? Don't you agree with my carefree view?"—she stuns him with her simple reply. "I don't know, but it seems like a strange mixture of the pessimistic and the carefree. I wonder if you aren't kidding yourself a bit."

Daisuke wants to change the subject, but in order to defend himself against the charge that he lacks a social function he repeats his earlier rationalization. "Only the man freed of economic necessity and following the natural inclinations of his spirit can engage in serious work." "That means," says Hiraoka firing back the statement in reverse form, "only a person in your position can perform holy labor! That is all the more reason why you should be working. Right, Michiyo?" Daisuke scratches his head at her affirmative reply. "We are back to where we started. It was wrong to have initiated this discussion in the first place." The argument is as good as lost.

This discussion with the Hiraokas reveals the surprising fragility of Daisuke's convictions. His tendency to project his personal problems on the world indicates that while, on the one hand, he is critical of the existing order, on the other, he expects it to indulge his whims (*amaeru*). This expectation also manifests itself in the imperviousness to shame he demonstrates at living off his father. I do not believe, however, that we can take Daisuke to task for his dependency wishes (*amae*) because he is not a person readily given to self-indulgence and because his posture as a social critic bars him from pursuing these wishes. The impulse remains, nonetheless; and, of more importance, it saves him from total alienation from his family and friends. In other words, his dependency wishes (*amae*) act as a safety valve in the maintenance of his precarious mental balance. Unfortunately for Daisuke, they are his only avenue to help.

They are evident, for example, in his criticism of his father's character. To Daisuke's thinking, his father is a prisoner of custom who had received and still adheres to "the kind of morality-centered education peculiar to men of the samurai class born before the [Meiji] Restoration." At the same time, he is also a shrewd businessman who has been corrupted over the years by an overwhelming appetite for money. He is unaware, of course, of "what had to be a considerable difference between his past and present selves. Daisuke felt that his father was either a sham of a samurai or an undiscerning fool. *He detested having to feel this way about him.*" [Doi's italics] Thus, while his sister-in-law is correct in believing Daisuke thinks his father a fool, she overlooks the heartache he experiences at this fact. If at all possible he wants to respect his father, and he suffers because he can neither respect nor communicate with him.

Daisuke's secret wishes are also evident in a passage describing his feeling when, having sought his sister-in-law's help, she sends money for the Hiraokas. "He replied immediately. He expressed his appreciation in the warmest language he knew. He never felt this way about his brother or his father, let alone the rest of the world. Lately he had not felt this way toward Umeko." Outward appearances notwithstanding, Daisuke is a lonely person. This may be related, as I suggested earlier, to the untimely death of his mother. Before the novel ends this lonely man will choose utter isolation by cutting himself off not only from his father and brother but also from his surrogate mother.

The Road to Catastrophe

When Daisuke calls on Michiyo to deliver the money, he learns that her relationship with her husband has deteriorated and that, since her illness, Hiraoka has taken to staying out late. He sympathizes with her because he too feels unhappy about the way in which he has become estranged from his friend ever since, no doubt, he had been bested in their argument. At any rate, Michiyo's news strikes a complex chord in Daisuke's psyche. She was the younger sister of a mutual friend of both men, and they had known her well. Shortly after the friend's sudden death from typhoid fever, Hiraoka revealed to Daisuke his desire to marry Michiyo, and Daisuke acted as go-between. The novel does not record any incidence of jealousy between the two men; and it appears that Daisuke enjoyed acting on behalf of his friend, "who was like a brother." We are told, however, that when Daisuke went to bid the couple farewell at the train station he noticed "an enviable look of triumph" flash behind Hiraoka's glasses. "Suddenly he felt hatred for his friend."

About the time Daisuke visits Michiyo he begins to suffer from vague feelings of anxiety. "Despite his highly nervous temperament, he had never really been subject to attacks of anxiety. He himself knew this to be the case." It does not occur to him that his relationship with the Hiraokas is the reason, and he attributes the change to physiological causes. As his previous statement made clear, he lives without specific goals, following the inclination of his tastes and enjoying life in dilettantish fashion. Yet lately he feels keenly "a lack of vitality." This may be the ennui to which he often refers, or it may be the unperceived aftereffect of his argument with Hiraoka. "He imagined that there was only one way to save himself from his effete existence." He mutters to himself, "I must see Michiyo again."

Although Michiyo is often in his thoughts, he is not yet in love or at least conscious of his love for her. On an earlier occasion when Umeko asks Daisuke if he has a favorite woman whom he would like to marry, "the name Michiyo came into his head." He finds this puzzling but attaches little significance to it. Consequently, when he speaks of seeing Michiyo again he says he *must* visit her rather than he would like to. How typical of Daisuke to take the simple, budding desire for love and transform it into a need to feel vitality.

Events move rapidly toward an ineluctable climax. Daisuke is maneuvered into a surprise interview with a potential bride and, as he comes under mounting family pressure to consent to the marriage, he begins to perceive the attraction Michiyo holds for him. He feels impelled to travel to escape this predicament, but when he visits Michiyo and sees that her circumstances have not improved in the slightest he abandons the trip in order to give her money. This, of course, makes him available for another round of marriage negotiations. He fears that to reject his father's choice may mean disinheritance and the loss of material support for which his only consolation will be Michiyo. Still, he does not consider taking her forcibly from Hiraoka. He pities her. He regrets having arranged her marriage. And at last he decides to speak to Hiraoka about being more considerate of his wife. The results are negligible, and he is left feeling more inadequate. Now at last he recognizes his lost love, as well as his two alternatives: either to take action and seize Michiyo for himself or to forget her forever and consent to his father's choice of a bride.

He is arduously slow in making his decision. After considerable mental suffering, he informs his sister-in-law and, in order "to fix his course so as to leave no margin for retreat," he calls Michiyo to his house and confesses his love. It is then that he sees his father who, predictably enough, announces: "I will not trouble with you again." Daisuke seeks Michiyo's company frequently, but as a secret liaison is untenable he decides to settle matters with Hiraoka. Hiraoka says, however, that Michiyo is sick in bed and he cannot surrender her until she has recovered. He severs relations with Daisuke and requests that he refrain from visiting his wife in his absence. The question of her actual transfer is left unsettled, and the situation takes an ominous turn when Hiraoka writes a letter to Daisuke's father. Daisuke is formally disinherited and ostracized from his family. As the novel concludes he leaves his house saying that he is going to search for work. He decides to ride the streetcar until the flames in his head burn themselves out.

In the preceding discussion we have covered the details of Daisuke's involvement with Michiyo. What we need to ask next is why it happens. The standard interpretation blames the spirit of chivalry out of which Daisuke originally relinquishes his claim on Michiyo as the source of his downfall. Both the passage in which he confronts Hiraoka with his demand for his wife—"I loved her before you"—and a subsequent one—"I was wrong to feel that true friendship lay in sacrificing myself in an attempt to grant your wishes"—are often cited as support. There are several reasons for believing, however, that these passages are not to be taken literally.

First, there is no evidence that Daisuke ever feels jealous. The novel does record as the one exception the hatred he feels at Hiraoka's triumphant look, but whether this constitutes jealousy is questionable. We would probably be more correct to interpret it as anger over the fact that, in spite of Daisuke's ministrations on Hiraoka's behalf, his friend is about to abandon him—and in a manner that suggests his happiness is entirely of his own making.

My second reason derives from the reply Michiyo gives to Daisuke's declaration of love. She cries and, quite naturally, demands to know why he rejected her originally. Were he truly in love, he should not have permitted her to marry unless, of course, his allegiance to his friend carried considerably greater weight. This suggests that Daisuke has latent homosexual tendencies and that his anger at Hiraoka's triumphant look is also related to them. To overstate the case a bit, Daisuke feels betrayed because marriage makes his friend inaccessible. I stated previously that the collapse of confidence in his father was the decisive cause of his neurotic tendencies. But the break with his friend also has its subtle influence and, in terms of the etiology of his illness, probably predates the break with the father.

My third reason concerns the circumstances surrounding the re-emergence of Daisuke's interest in Michiyo. Although it is true that he had always liked her, his fondness never amounts to love. The strong sympathy that he feels for her, however, combines with a vague sense of hostility that he entertained toward her husband over the last three years. His defeat in the argument, moreover, heightens this sense, and it has the effect of convincing him that Michiyo is as much a victim of Hiraoka's misadventures as he is. Consequently I believe we can say with considerable certainty that he desires Michiyo as a means of retaliation. This idea does not surface until late in the novel because it is an objectionable thought and because Daisuke persuades himself that Hiraoka is at fault for having failed to love his wife. It operates on an unconscious level, nonetheless; and the word "revenge" slips from Daisuke's lips when he proposes to Michiyo. Her denials to the contrary, he wonders if she will not revenge herself upon him for not proposing three years earlier. At this point his speech becomes jumbled and illogical, but it helps to know that he is projecting on Michiyo the vindictive feelings he cannot direct to Hiraoka.

My fourth reason for questioning the standard interpretation stems from the fact that Daisuke uses his new alliance with Michiyo as a means to break off the marriage negotiations and declare his economic independence. We might even say he utilizes her as a means to an end, and there is no better proof of this than his view of her as his consolation. At the same time, we probably must also recognize that, seen from his subjective viewpoint, he genuinely loves Michiyo. Michiyo makes it possible for him to abandon the role of onlooker and for the first time in his life to leap into the real world. He has lost, however, his former self-confidence as well as the channel to the gratification of his hidden dependency wishes (*amae*). Can he now be denied the right to see her and survive? The emotional crisis he thought was resolved returns to haunt him, and this time he must do battle stripped of his defenses. The final scene of the novel suggests that Daisuke's emotional state is aggravated enough that he may be on the verge of psychosis. We are kept from knowing, as the title *Sorekara* implies, what will happen thereafter.

Chapter Five

MON (The Gate)

1910

Synopsis

This novel is a somber portrait of Nonaka Sōsuke and his wife, Oyone. Although they care for one another with a singular devotion, their love is insufficient to dispel the poverty and boredom that characterize their life in a drab rented house at the foot of a cliff. They do not expect a better life because they live in the shadow of their past.

As a well-to-do student at Kyoto University, Sōsuke's success in life seemed assured. But then he committed what the novel refers to as an indiscretion with Oyone, who was at the time the common-law wife of his best friend and classmate, Yasui. This act ended the college careers of both men and altered their lives forever. Sōsuke, unable to return home and face his father, eloped with Oyone to Hiroshima where he found work as a petty civil servant. Yasui, it was rumored, disappeared to Mongolia.

Six years have elapsed and now the Nonakas live in Tokyo, Sōsuke's birthplace. His parents are dead, and all that remains of his mismanaged inheritance is an antique folding screen. Sōsuke and Oyone no longer discuss their past, but there are occasions when each is privately reminded of the wrong he did.

Such is the case this fall when the question of financing a college education for Sōsuke's younger brother, Koroku, arises. An uncle, Mr. Saeki, raised the boy in the years of Sōsuke's absence from Tokyo, but he has died and his wife is unwilling to provide for her nephew beyond higher school. Although Sōsuke cannot handle the expense, he feels obligated to give his brother the education he himself wasted.

After considerable procrastination, it is decided that Koroku will room with the Nonakas until enough money can be saved to send him back to school. This arrangement satisfies neither Koroku nor Oyone. It is hard enough for Oyone to bear the burden of being unable to give her husband the children that, according to one fortuneteller, fate denies her for having deserted Yasui. Now she must share her small home with Koroku who blames her for the misfortune that has befallen his brother and now him.

Autumn passes quietly with the tedium of the Nonakas' lives interrupted only by minor crises. The heirloom screen is sold to provide Sōsuke a badly needed new pair of shoes; Oyone suffers from a brief but ominous attack of chest pains; Koroku takes to tippling.

One night a noise is heard outside the house and, the following morning, Sōsuke finds a small lacquer box discarded in the backyard. It belongs to the landlord, Mr. Sakai; a robber had broken into his house atop the cliff. Sōsuke avails himself of this opportunity to become friends with his landlord. The Sakai household rings with the laughter of children and is filled with nice possessions. On a second visit, Sōsuke learns that his heirloom screen is now, ironically, in Sakai's possession. Sakai purchased it from the same pawnbroker who had bought it from Oyone at half the price.

The Nonakas have little to look forward to in the new year until one evening when Sakai unexpectedly offers to pay for Koroku's education in exchange for the boy's services as a live-in servant. Koroku had made a favorable impression when he called on Sakai to deliver traditional New Year greetings.

Sakai mentions that he too has a younger brother, an "adventurer" whose grandiose business schemes on the Asian continent cause him much concern. The brother has recently returned from Mongolia and is expected at the house. Sakai invites Sōsuke to meet his brother and his brother's friend, a man known to him only by the name Yasui.

Sōsuke is tormented by the thought of Yasui's reappearance. He becomes sullen and feels nauseated; and, on the night of the proposed meeting, he avoids returning home and tries to find solace in drink. He remembers how an old classmate once practiced

Zen meditation and, although he always scoffed at the idea, he takes leave from work and goes on a retreat to a Zen temple in hope of finding deliverance from anxiety. Oyone is distressed by the sudden change in her husband's behavior, but Sōsuke will not tell her the real reason for his trip.

At the temple Sōsuke is assigned a kōan *as a topic for meditation. It serves only to intensify his mental anguish, and he is chided by the Zen master for failing to apply himself. Ten days later, as he leaves, tired and disheveled, he is no nearer enlightenment or ease of mind.*

He calls on Sakai who reports that his brother and friend have returned to Mongolia; it appears that Yasui did not learn that Sōsuke lived nearby. Sōsuke returns to work and, despite a recent shake-up in the bureaucracy, retains his job and even receives a raise. Life improves for the moment, but he knows that someday Yasui may appear to confront him with his past. When, in the final lines of the novel, Oyone expresses her joy at the return of spring, he adds fatalistically that winter cannot be far behind.

Sōsuke's Life Story

The novel opens with desultory conversation between Sōsuke and his wife, Oyone. Sōsuke, who is catching the warm autumn sun on the veranda, calls into the house to ask Oyone how to write a simple Japanese character. He remarks that recently his memory had failed him when it came to writing another, far simpler, word.[13] "Once you think you've written even the easiest of characters incorrectly, you can look at it all day and never find your mistake." His tone is serious, and he wants to know if ever Oyone had a similar experience. She wonders if "there is something wrong" with her husband; and he concurs—"Perhaps it is my nervous condition again." Sōsuke is exhibiting, in fact, symptoms of depersonalization. Simple characters that should be permanently fixed in his mind have lost their sense of reality, and he has begun to ruminate about them in a compulsive manner. This suggests that he is experiencing some type of deep inner turmoil.

It will be our task to elucidate the source of his problem, but even Sōsuke is aware that in comparison with the past he is a changed man. The sight of his younger brother, Koroku, never fails to evoke the memory of "his old self, alive again and active before his own eyes." As a university student and the son of a comparatively wealthy Tokyo man, Sōsuke had been free to go when and where he pleased, "holding his head high for the world to see and projecting an image of a talented young man thoroughly *an courant* in thought and dress." He was forced to withdraw from school during his second year, however, for committing "an indiscretion" with Oyone, who was married at the time to Sōsuke's best friend and classmate, Yasui. Unable to return home and face his family, Sōsuke eloped with Oyone to Hiroshima, where he found employment. Six months later his father died. (His mother had passed away six years earlier.) Sōsuke was transferred to Fukuoka half a year later, and after two years in Fukuoka a friend arranged for him to return to Tokyo.

The novel provides scant detail about the events leading to Sōsuke's indiscretion, but on the basis of the available facts it appears that his excessive friendliness with Oyone was the source of his downfall. One expects a friend to exercise restraint in his relations with one's wife; and, initially, Sōsuke was circumspect in his conduct toward Oyone. Eventually the two became such friends that they ceased to find it strange to be found alone in one another's company. Yasui, too, did not seem to question their familiarity, and there is even a passage that suggests he welcomed it. We might note in this connection that the two men were close friends and constant companions. There was a brief cooling of relations when Yasui married without informing Sōsuke, but in no time their friendship resumed with renewed vigor and Oyone's presence added new sparkle to an old association. Oyone's wifely status did perhaps make it easier for Sōsuke to become friends with her, and no doubt Yasui encouraged their friendship as a means of bolstering his own with Sōsuke. One can also imagine that, since the death of his mother, Sōsuke had grown hungry for the warmth and comfort of a real home and that this need drew him to Oyone. We can say, therefore, that the relationship between the two men was probably latently homosexual and that there was a strong oedipal coloring to Sōsuke's friendship with Oyone. This is not to suggest anything abnormal—for as far as the three were concerned their relationship was extremely guileless. It is precisely because of this innocence that they were incapable of foreseeing the danger inherent in the relationship.

Of course, Sōsuke's indiscretion signaled an abrupt end to this blissful triangle. Oyone deserted Yasui for Sōsuke, and Yasui abandoned his plans to finish his education. So it was that Sōsuke's life was irrevocably altered. He lives in fear of what people will say; and, as a result, his aunt remarks upon his return to Tokyo that he has aged out of proportion to the three years of his absence. The change is not merely psychological either. Sōsuke also suffered financial reverses as a result of the death of his father. He had entrusted disposal of his father's house to his uncle who, in Sōsuke's absence, arbitrarily reinvested the money in property that was subsequently destroyed by fire. At the time Sōsuke hesitated to complain about the investment because he was indebted to his uncle for assuming responsibility for Koroku's upbringing and education. Now that the uncle is dead, there is no one with whom he can remonstrate.

All that remains to him is his life with Oyone. They close their eyes to society and live solely for themselves. "Each really needed only the other; and so long as they had each other they were sufficient unto themselves." To an outsider their marriage may seem happy, but it does not escape a touch of loneliness. It provides them with warmth and protection against the cruel glare of society, but it does nothing to insure their future success and personal growth. They are not blessed with children; and this fact, by making their loneliness more poignant, perhaps best symbolizes the inner emptiness of their lives.

Had it not been for external pressures, they might have remained unaware of this inadequacy. At the point the novel begins, however, Sōsuke is confronted once again with the worrisome question of how to raise money for his brother's education. In the past he had left the matter in his uncle's hands, but his aunt has announced that, owing to changes in her financial situation since her husband's death, she can no longer provide for Koroku. Of course, it is burdensome enough for Sōsuke to provide for himself and Oyone, and he is unable to absorb the additional expense. As luck would have it, the problem is solved when Sōsuke happens to become friends with his landlord and the landlord offers to pay for the boy's education in exchange for Koroku's services as a houseboy. The offer is, however, accompanied by a startling bit of news, namely, that in the near future the landlord anticipates a visit from his younger brother and the brother's companion, a man named Yasui. Sōsuke is shaken to the core. Raising money for his brother's education may have been a source of aggravation, but it did not leave him panic-stricken. Yasui's reappearance is, however, an event of a magnitude sufficient to destroy the emotional equilibrium that he has struggled so long and hard to achieve.

Consciousness of Guilt

As I have already stated, the peculiar set of circumstances surrounding the marriage of this couple leads Sōsuke and Oyone to think of themselves as persons dead to society. It is for this reason that Sōsuke rejects the aspirations of his youth and contents himself with the life of a social outcast. Likewise, Oyone attributes their childlessness to their sin. She has had three pregnancies—the first ending in a miscarriage; the second in a premature delivery and death; and the third in a stillbirth. After the third failure she decides to consult a fortuneteller, who says she will remain childless. "You have committed a grievous sin against another person, and that sin is still working itself out." As time passes Oyone becomes convinced of the validity of this explanation. The shadow of the past hovers over the lives of this couple, and in this sense they can be described as persons suffering from a consciousness of having done wrong. There is doubt in my mind, however, as to the exact manner in which they view their wrongdoing. Wouldn't it be more precise to say that they suffer from a consciousness of punishment rather than a consciousness of guilt? This is what Sōsuke indicates when he says, "We have no right to look forward to better times." Not only have they lost their rights and privileges as human beings, they have also lost the right to protest against that loss.

In contrast to this vivid sense of punishment, their awareness concerning the misdeed that led to the punishment is much less sharply defined. The novel records that "thinking back to that time, Sōsuke mused about how, if the cycle of the seasons had only come to a halt, if he and Oyone had in that instant turned to stone, vast suffering could have been forestalled." Or, again, "A great wind had swept in from nowhere and blown them off their feet." These passages suggest that the two look upon their misdeed as the handiwork of blind forces of nature and upon themselves as nature's victims. They know of course that their behavior was indefensible, but as they conceive of their sin in a heteronomous sense, i.e., in the sense that "society was unmercifully holding them to account," one must say that their awareness of having committed an immoral act is most superficial. Consequently, "they themselves, before feeling any guilt, wondered whether they had been entirely sane at the time. Prior to suffering the shame of feeling immoral, they were amazed at themselves for having been irrational." Seen in this light their consciousness of wrongdoing is all the more specious. Perhaps the exaggerated innocence that characterized their early friendship subsequently made it difficult for them to believe that they were the authors of their own indiscretion. No doubt they thought of themselves as completely innocent. "They were filled with bitterness because fate had whimsically taken two *innocent* people and dropped them into a trap for the sake of its own amusement." [Doi's italics] Although they did not entirely escape feeling guilty, "by the time the full weight of their misdeed had fallen squarely on their shoulders they had ceased to suffer from further pangs of conscience." We can assume that Sōsuke and Oyone had done their utmost to avoid a confrontation with their sense of guilt.

While, logically speaking, it may be contradictory to speak of deserving punishment while lacking a sense of wrongdoing, this is a possibility in the psychological realm. What is required, however, is the expenditure of an appreciable amount of psychological effort because the ego is continually caught in the task of denying its own guilt and trying to avoid any situation that threatens to reawaken it. In this respect the following passage is most interesting. "The demands of fate and perseverance dominated at all times. Only on rare occasions did a glimmer of hope, or of the future, shine into their lives. They said little about the past. It even seemed as though they had mutually agreed to exclude it from their conversations." Their lack of "hope or of a future" stems from a consciousness of punishment. Their reluctance to discuss the past stems, however, from a desire not to be reminded of their sin. They avoid not only what others think but also what they themselves think.

I stated earlier that Sōsuke's symptoms of depersonalization are indicative of inner turmoil. It is apparent now that the source of his problem is none other than his denial of his consciousness of guilt. We can imagine that the symptoms became manifest once his psychological efforts at denial expanded to include not only the consciousness itself but all factors directly or indirectly related to it. Similarly, we can now appreciate the full significance of Yasui's impending arrival. It becomes, literally, physically impossible for Sōsuke to continue to utilize the denial of guilt as a psychological defense mechanism, and his emotional equilibrium is destroyed. He does not avail himself of the opportunity to confront his past, however. He becomes preoccupied with the feeling that he is being victimized. "*He did not consider himself so strong* that an opponent need take advantage of happenstance and, slipping up from behind, fell him with fancy legwork." [Doi's italics] This statement reveals the tremendous resentment Sōsuke feels at having his vulnerability exposed. "Under the massive burdens that weighed him down, he could think only of strategies for getting free. As a result, he separated effect from cause, the burdens themselves from the wrongdoing that had created those burdens."

Overcome by deep frustration he takes leave from work and retreats to a Zen temple in Kamakura. His ten-day stay reminds him once again that he is a lost soul, and his meditations are often wildly distracted by the thought that Yasui has heard accidentally of his existence from the landlord. He feels considerably relieved when a day or two after he returns home he calls on the landlord and learns in a roundabout way that his fears are groundless. Nonetheless, "something told him that he was destined to experience this anxiety any number of times and in varying degrees of intensity. Such was the will of Heaven, and it would be his part to try to escape it." No doubt Sōsuke is partially aware of his own desperate efforts to deny his sense of guilt.

The Unbelieving

One reason for Sōsuke's lack of success at Zen meditation may be the brevity of his stay at the temple; another is the shortsightedness of his objective—he is interested only in sparing himself from further pain. So that upon being presented with the *kōan* "What was your true Face prior to the birth of your parents?" "he couldn't help but feel that it was of no relevance whatsoever. It was as if he had come in order to be cured of a stomach ache and been given a hard mathematics problem with the instruction that he ponder it a while." Apparently, Sōsuke lacks the prerequisite attitude for sitting at Zen. He is not interested in seeking release from the worldly desires that Buddhism holds to be the source of human travail. Although he has rejected the world, he has no desire to renounce it in the sense that a Buddhist takes the tonsure to become a monk. He is, in a word, an unbeliever.

This is not to say that he never considers the subject of religion. On one occasion he asks Oyone if she has ever experienced the desire to believe in religion. The subject is not developed, and it is not clear whether this is because of their lack of interest or the lack of religious alternatives in their area. "The two of them had lived without ever sitting in the pew of a church or passing under the gate of a temple. Whatever degree of peace they had managed to find came from the healing qualities of time's passing, one of the benefits of Nature. And if, from time to time, they heard a remote voice reminding them of their past, the voice was too dim, too far away, too removed from lust or greed to merit our branding it 'torment' or 'terror.' In the end, because they lacked the religious belief they would have needed to recognize either God or Buddha, they lived committed only to their mutual well-being. Holding each other tightly, they drew a protective circle around themselves. Their daily life became stabilized in a wistful loneliness, and in this mood they found a comforting sorrow. As neither of them had much acquaintance with art or philosophy, they partook of this sorrow without ever developing the pride of the intellectual who congratulates himself upon his self-awareness of his own adversity. In this sense, their experience was far purer than that of poets and men of letters who have found themselves in similar straits."

This passage is an apt description of this couple's life until the time Sōsuke learns of Yasui's arrival. The news produces a complete change in his emotional state, and he is thrown into extreme anxiety. Curiously enough, he chooses to conceal from Oyone the reason for his unhappiness. Unable to turn to a higher being, he should have relied upon his wife for help, and the fact that he does not suggests that the belief they have constructed of their togetherness is too fragile to withstand his latest attack of anxiety. It is at this point that he recalls how a friend had once practiced Zen meditation and, in a fit of desperation, he decides to attempt it. His efforts prove unsuccessful; and one imagines the results would have been much the same had he sought the solace of Christianity instead. I say this because the teachings of Christianity are even further removed from Sōsuke's needs than those of Buddhism.

As a matter of fact, Sōsuke shares his anxiety with the great mass of men and women who do not possess a religious belief. In this respect he represents not only modern Japanese but also many contemporary Westerners. For these people religion no longer has meaning; it is a vestige of the past that fails to speak to their problems. They look elsewhere for salvation from pain and anxiety, and no doubt this accounts for the popularity of psychoanalysis among certain sectors of western society. Because psychoanalysis is very much the product of our unbelieving age, it is all the more capable of appreciating their suffering; and I suspect that, while it may not attract widespread interest in Japan at present, in time more Japanese will require a type of world view such as psychoanalysis affords. I say this because it is meaningless to suggest to people like Sōsuke that they should be less selfish to worldly desires when they are so emotionally exhausted they can no longer recognize their own self-interest. Or to tell them to repent their sins when they are desperately trying to deny them. These people require help in analyzing and gaining insight into the causes of their pain and anxiety.

Staying within the limits of the novel as much as possible, I should like to consider the hypothetical question of what might have happened had Sōsuke been under the care of an analyst. During his stay at the temple Sōsuke comes to realize that he is a helpless person. "This was a revelation—a discovery powerful enough to destroy the last remnant of his self-respect." No doubt he would experience the same feeling in undergoing psychoanalysis. Again, at the conclusion of his stay, he says, "I had arrived at the gate and waited to have it opened. But the gatekeeper was on the other side and did not appear when I knocked. I heard only a voice that called, 'Knocking will achieve nothing. Open it for yourself.'" Here we have the clearest expression of his feelings of dependence. In the psychoanalytic setting these feelings would be transferred to the analyst and made conscious. If Sōsuke succeeded in developing a genuine trust for the analyst, he would probably come to accept the analyst's interpretation not only of the reasons for his longstanding avoidance of his consciousness of guilt but also of the nature of the feelings that had originally tempted him to commit the indiscretion. As a final step he might also realize that, although he may have committed the act in ignorance, the acceptance of guilt provides the only path open to him if he is to live as a human being.

It is at this point that he must confront a subject that has been very foreign to him, namely, the question of religion. It is no easy task to accept one's guilt, and Sōsuke may earnestly feel the need to seek forgiveness of his sins or to discover a state beyond sin such as implied by his *kōan.* In the event, however, that the analyst is also an unbeliever, Sōsuke may, by identifying with him, choose to accept life by regarding his sin as an inescapable fate and its absolution or transcendence as an illusion. This is a kind of modern stoicism which, as a matter of principle, repudiates consideration of the metaphysical.

If he should go one step further and dismiss the analyst's stoicism as illusory, he may in his despair at being unable to escape his newly found consciousness of guilt attempt suicide. It is impossible to predict which path he might choose. Whichever it is—whether Sōsuke accepts or rejects religion—he must at least seriously consider the question. Even if he is one of the unbelieving.

Chapter Six

HIGAN SUGI MADE (By After the Equinox)

1912

Synopsis

This novel consists of six loosely knit stories which Sōseki, after a prolonged illness, vowed to complete by no later than after the spring equinox of 1912.

In "After the Bath," Tagawa Keitarō is discouraged by his lack of success at finding employment since his graduation from the university. An insatiably curious and romantic youth, he daydreams of exotic jobs in faraway places.

He would like to have led the life of his fellow lodger Morimoto. Morimoto was adventurous in his youth and has an inexhaustible supply of tales on every subject from antimony prospecting to seal hunting. One morning when the two men meet at the public bath, Morimoto says that the wanderlust is upon him again; and it is not long before he disappears without a trace. Later he writes Keitarō from Port Arthur, Manchuria, where he has found work in an amusement park. He wills to him the few possessions he left behind. These include an unusual walking stick in the shape of a snake with a gaping mouth. It stands in the hallway rack.

In "Streetcar Stop," Keitarō calls on his former classmate, Sunaga Ichizō. Sunaga lives with his mother in a quiet neighborhood that retains the atmosphere of old Edo. Although graduated with honors and given an immediate offer of a job, he has no desire to work and is content to live on an income left by his late father.

Keitarō notices a young woman disappear into the house seconds ahead of his arrival. His curiosity is aroused and he wonders if she is Sunaga's fiancé, but Sunaga is alone in his study and talks of other matters. He has promised to introduce Keitarō to his uncle-in-law, a businessman named Taguchi Yosaku. When he presses Keitarō as to the type of work he desires, Keitarō admits to a secret longing to be a detective.

Taguchi is too busy to see Keitarō on the two occasions when the boy calls for an interview. Out of frustration and anxiety, Keitarō decides to consult a fortuneteller, who says that the resolution of his problem lies in the possession of "something that is his but not his, that is long but also short, that is opening at the same time it closes." Keitarō is convinced that the object is Morimoto's walking stick and he surreptitiously removes it to his room.

He carries it as a talisman when, finally having seen Taguchi, he is given an "assignment." He is to observe the movements of a man scheduled to appear at the Ogawa-machi streetcar stop. He will be wearing a pepper-and-salt tweed overcoat and has a mole over the bridge of his nose. Two hours pass before the suspect arrives and approaches a young woman. Keitarō follows the couple into a restaurant where he is able to overhear only snatches of their conversation; his attempt to follow the man home also proves unsuccessful.

In "The Report," he is crestfallen at being unable to report in detail. When he suggests that approaching the suspect directly might be a more expedient means of obtaining information, Taguchi agrees and gives him a letter of introduction. He warns that the man may not be very accessible and, on his first visit, Keitarō is turned away for the curious reason that "it is raining." He is received the following day, a clear day, and his questions are answered with great equanimity. The man is Taguchi's brother-in-law, Matsumoto Tsuneo; the woman was Taguchi's daughter, Chiyoko. What seemed like a lovers' rendevous to Keitarō was nothing more than an uncle taking his niece out to dinner. Keitarō has been the butt of the latest of the many practical jokes Taguchi likes to play on people.

Taguchi rewards him with a position as his personal aide, and through it Keitarō gains access to the private lives of Matsumoto, Sunaga, and Chiyoko. From Chiyoko he hears the sad, elegaic story of "The Day It Rained." Matsumoto has refused to see callers on inclement days ever since one rainy night the previous autumn when his youngest daughter, Yoiko, died suddenly and inexplicably. From "Sunaga's Story" and "Matsumoto's Story," Keitarō learns why Sunaga will never marry his cousin Chiyoko.

Although the cousins had been childhood playmates and Chiyoko possessed a charming personality, Sunaga gradually lost interest in her during his years in higher school. This was as matters should be, he thought. He was startled one day when his mother announced that Chiyoko had been promised to him since birth and, breaking into tears, begged him to marry her. He procrastinated and even expressed to his Uncle Taguchi his hope that a more suitable candidate be found for Chiyoko.

In his third year of college, he and his mother went to visit the Taguchis at their summer resort in Kamakura. There he found Chiyoko in the company of a dashing young man named Takagi. He became wildly jealous and returned to Tokyo. So great was his jealousy that he fantasized about bashing a heavy paperweight into Takagi's skull. When Chiyoko accompanied his mother home two days later, he could not resist asking about him. Chiyoko was deeply hurt by Sunaga's expectation that she be loyal to him without so much as a promise of love or marriage. She called him a coward and a fool.

Shortly before Sunaga's graduation, his mother asked Matsumoto to make a final attempt to persuade her son to marry Chiyoko. Sunaga could not fathom the reason for her insistence and he accused his uncle of withholding information from him. Matsumoto decided then and there to tell him the truth: Sunaga was the illegitimate son of his father and a family maid, Oyumi. Mrs. Sunaga had raised him as her own child; marriage with Chiyoko was her means of tying him closer to the family.

Sōseki notes in "The Conclusion" that Keitarō's career consists of the passive observation of the lives of others. It is only when he appears with his lucky stick that he commands the reader's attention. It is a bit part that, while humorous, fails to answer the question of how he is to live his own life.

Keitarō's Curiosity

It is the overweening curiosity of the central character, Tagawa Keitarō, that serves as the driving force of this novel. He is the type of person who "not only reveled in fantasies of faraway people and places but also, right in his own city, imagined that everyone—the woman on the train, the man on the street—was hiding something extraordinarily exotic under his cape or up his sleeve." The strange tales he hears from his fellow lodger, Morimoto, are a source of constant entertainment for him. Then one day Morimoto disappears abruptly from the rooming house. This is the sort of incident calculated to arouse Keitarō's curiosity, or so it would appear until he is approached by the landlord. Morimoto is six months behind in his rent, and the landlord would like to know of his tenant's whereabouts. "Keitarō was greatly annoyed at the way the landlord seemed to suggest that he, as a friend of the missing party, was involved in this distasteful act." Sometime later Keitarō receives a cordial letter from Morimoto. He does not, however, inform the landlord; nor does he answer it immediately. It takes several attempts to compose an acceptable reply, and the manner in which he drops the letter into the mailbox suggests that he is ridding himself of a nuisance.

Although Keitarō enjoys prying into people's secrets, he fears any injury that may attach to him as a result. He would like to be a detective were it not for the social stigma associated with the work. He says, "I am merely a student of humanity. It is my wish to bring a sense of awe to the observation of the wild schemes men devise in their attempt to navigate through the dark night of life." These are the words, of course, of a man who, rather than live his own life, prefers to experience it vicariously through, for example, reading novels. Indeed they may well be the words of a reader of novels who, while perhaps capable of turning out a book or two, could never himself figure in one as the central character. It might prove interesting to ask ourselves why Sōseki chooses a bystander like Keitarō to be his protagonist. This question is especially curious in view of his avowal in the preface of the novel that he will write in a style that is uniquely his own.[14]

Soon he has us following Keitarō's trail. Sunaga Ichizō, Keitarō's friend from college, has approached his uncle, a businessman named Taguchi, concerning the possibility of a job for his ex-classmate. Word arrives to the effect that Keitarō is to spy on the movements of a certain person scheduled to appear at a specific time and place; and, as a consequence, he becomes party to a rendezvous between a middle-aged man and a young woman. When he reports to Taguchi the following day and is unable to offer explicit answers to Taguchi's many questions, he gives the following defense. "I am terribly sorry I cannot make better sense for you. I'm afraid my mind is not sharp enough to discern in so short a time the kinds of personal information you desire. Excuse my brashness, but it seems to me that it would be far less onerous and far more accurate to confront these people outright rather than engaging in the petty trick of following them." Surprisingly enough, Taguchi is impressed by this argument and he apologizes for his ignorance—Sunaga had led him to believe that Keitarō wished to be a detective. Next he asks if the boy has the courage to meet his suspects, and when Keitarō agrees he gives him a letter of introduction to a Mr. Matsumoto. Although Keitarō is sent away on his first visit to the Matsumoto residence for the strange reason that "it is raining," the man does receive him when the skies clear on the following day. Keitarō learns that Matsumoto is Taguchi's brother-in-law and the young woman is Taguchi's daughter.

Keitarō has been the victim of a practical joke and, while this cruel initiation guarantees him the kind of position he has always wanted, his budding curiosity suffers a definite setback. Whereas it had been frustrated in the case of Morimoto's disappearance by fear of implication, this time it brings him to shame by letting him become overly enthusiastic about an assignment that is, doubtless, tailor-made to his tastes. As Keitarō expects, Taguchi has a hearty laugh at his expense when the two meet again. Taguchi does not, however, look down upon the boy. He asks him not to be angry—"It was not out of spite"—and, by offering full-time employment, saves him from serious loss of face. It appears that after this episode Keitarō decides to hold a tighter rein on his curiosity—to the extent, in fact, that he forgets to investigate why Matsumoto does not receive callers on rainy days. It is not until later when he happens to meet Taguchi's daughter, Chiyoko, that he learns the reason.

There is a second matter that arouses his curiosity, and this one he is unable to forget. It dates to one day shortly before he entered Taguchi's employ and it concerns a young woman he had seen entering Sunaga's house moments ahead of himself. Keitarō fantasizes that his friend is romantically involved with this woman. Later he realizes that she is the same person that he had observed in Matsumoto's company, namely, Chiyoko. Even after Keitarō goes to work for Taguchi, "his thoughts were dominated by his initial association of Chiyoko with Sunaga. He tended always to think of them as a couple." He cannot find anyone to support him in this belief and, when he hears that Chiyoko has been made an offer of marriage from another quarter, he is stirred by a desire to confront his friend with the news. Thus, one Sunday he invites Sunaga for a walk in the country and uses the occasion to discuss for the first time the status of his friend's relationship with Chiyoko.

"Sunaga's story was far longer than Keitarō anticipated." Instead of simply stating his intention not to marry his cousin, Sunaga begins his story with the indelible impression his father's death made upon him as a child. He discusses both the nature of his parents' personalities and the nature of his relationship with Chiyoko, his childhood playmate. It is his opinion that such a marriage would prove unhappy despite his mother's longstanding wish to wed the two. He has observed that neither Chiyoko nor her parents seem enthusiastic about the arrangement, and he believes that his personality is by nature ill-suited to hers. This last point defies Keitarō's comprehension because his overall impression of Sunaga's story is one of partiality for Chiyoko. The question that emerges in his mind is whether Sunaga does or does not love Chiyoko. Assuredly Sunaga had loved her in the past, but Keitarō wants to know the present status of their relationship. It is at this juncture that Sunaga relates the story of the tragic confrontation that occurred during the summer vacation of his junior year in college.

By the time Sunaga finishes, Keitarō realizes that the situation is far more critical and complicated than he imagined. Moreover, he feels a strong urge to bring the two together, and this is probably why he visits Matsumoto. He is told that the problem will not yield to his intervention, and he learns for the first time the facts surrounding the secret of Sunaga's parentage. Sunaga is the child of a servant who had been employed in his father's household. Shortly after his birth his present mother assumed responsibility for his upbringing. Matsumoto disclosed this fact to his nephew several months before his graduation and, of course, Sunaga is fully cognizant of it at the time he tells his story. He elects, however, to hint at—or we might even say, to conceal—the truth. Were Keitarō more perceptive he might have detected this bit of camouflage work, but his mind is preoccupied with his questions about Sunaga's relationship to Chiyoko, and Sunaga's hints pass unnoticed. We can also imagine that Sunaga perceives the focal point of his friend's interest and uses it as a means to conceal his great secret.

One is tempted to wonder if Keitarō ever understands Sunaga's problem fully. It is almost too involved for his simplistic mind. The novel makes, in fact, the following comment in the last chapter. "Keitarō's adventures began with a story and they end in a story. He had wanted to discover the world, and at first it seemed far, far away. Nowadays it is right at the tip of his nose. But he still resembles a spectator who can neither mount the stage nor perform in the drama. His role is confined to that of a narrator who holds a crystal set receiver to his ear and listens to the voice of the world . . . All in all, he owes his newly acquired knowledge of life to the functioning of his tympanic membrane. The divers tales that he has heard from Morimoto to Matsumoto have come to a crashing halt. At first he was only slightly and impersonally involved; now they touch upon the very heart of his existence. Never once has he taken part, however. This is the source of both his dissatisfaction and his good fortune."

It is apparent that, Sōseki's treatment notwithstanding, Keitarō is the type of person who lacks any of the qualifications vital to the central character of a novel. It is true that he draws out the personalities of the other principal characters, but in that sense he is nothing more than a literary device. He has, nonetheless, the temerity to act as though he were the central character, and critics do not seem to hold any specific opinions as to why Sōseki chooses to have him act in this fashion. In general, they have failed to address themselves to this question or they have attributed it either to artistic license or to poor construction. I cannot help thinking that Keitarō is, however, the product of the author's secret design for, through him, Sōseki seems to be drawing a portrait of his typical reader. Perhaps the reason that many readers find this character mildly irritating may be that he reminds them of themselves.

In other words, people are attracted to novels usually out of curiosity, and the author has no guarantee that they will endeavor genuinely to understand what he is trying to communicate to them. So that, just as Sunaga withholds vital information from Keitarō, Sōseki may also have been tempted to conceal his own secrets rather than face the considerable likelihood of being misunderstood. I would speculate that it is this type of ambivalence on Sōseki's part that gives rise to Keitarō's creation. No doubt he represents Sōseki's outward commitment to placate his audience with, as he states in his preface, something entertaining. But he also embodies a biting satirization of people who read solely for purposes of amusement. I believe it would be safe to say that this satirization is directed—and quite pointedly—not only at Sōseki's contemporaries but also at subsequent generations and the many Sōseki specialists they have produced. I cannot help feeling that Sōseki has the last laugh at people who never understand his novels no matter how many times they read them. Of their lack of comprehension, he says, "That is the source of both [their] dissatisfaction and [their] good fortune."

Sunaga's Secret

Sunaga withholds his secret because he feels Keitarō is unworthy of it. We can also imagine that he prefers to avoid reawakening the painful memories that accompany it. To a great extent it was a secret for him too. His earliest premonition dated back to his childhood when his dying father had turned to him and said, "Ichizō, you realize that once I am gone it is your mother who will have to care for you. You would be wise to behave a little better because she will not tolerate much mischief." Since the boy naturally assumed his mother would care for him, "he thought it strange that his father would make a point of the matter." On the day of the funeral, moreover, his grief-stricken mother whispered, "You need not worry. I will love you even though your father is no longer with us." The subject is never raised again, but as Sunaga grows older these words come to cast a shadow over his feelings about his parents. He cannot explain why he attaches special significance to them. "At times he wanted to ask his mother, but invariably he lost his nerve when it came to facing her." A voice inside whispers "That would be the end. Parent and child would go their disparate ways and never recapture their present sense of togetherness."

It is fairly easy to guess the nature of Sunaga's misgivings. A child will often fantasize that it is adopted when in fact it is not; and an adopted child often detects the truth of its parentage despite the foster parents' most careful efforts at concealment. It will pick up the slightest clue, and the clue is frequently a slip of the tongue made by an outsider. It is known that such slips have the effect of reinforcing the child's instinctive perception of an air of artificiality surrounding its relationship with its parents. This is perhaps the reason Sunaga attaches undue importance to his parents' remarks. This impression continues to grow, moreover, because his relationship with his mother is built upon a psychological uncertainty that belies the heightened degree of intimacy they have shared subsequent to his father's death. Sunaga's fear of the breach that may result from a candid discussion of his misgivings speaks only too well of the precariousness of their relationship. In addition, "he compiled over the years secret and detailed researches concerning the similarities and dissimilarities between himself and his mother." The discovery of similarity is a source of joy for him, but it also points to his need to prove to himself that he is his mother's child.

"He was not a docile child. His father's deathbed admonitions became just so many words, and he often defied his mother. Even after he grew older and he came to appreciate the need to be considerate of this widowed woman, he still would not be truly obedient." One imagines that this rebelliousness is a reflection of a complex set of emotions. Thus while, on the one hand, his suggestion that his mother will, as a mother, love him "no matter what" may appear to be an expression of the confidence he feels at being able to impose upon her love and patience (*amaeru*), it could also represent his unconscious need to test the limits of their relationship. It is possible to believe that his inability to discuss his misgivings with her makes this need more compulsive.

This does not mean, however, that the two live in a state of perpetual disharmony. It appears that as Sunaga grows older his rebelliousness subsides and outwardly they live in peace and happiness. On major issues, however, he will not accede to his mother's wishes, as is apparent from his refusal to seek employment. Unlike Keitarō, he had been offered a position even before graduating from the university. His rejection of the offer is not, it appears, a premeditated act of rebellion because he does experience guilt at the anxiety he knows he is causing his mother. The sincerity of his guilt is, however, open to question. "I could insist on my own way," he says, "as long as I appeared to be sorry." Although he cites a number of contradictory reasons for not working, none is very convincing: he is too introverted, too lacking in conviction; he is selfish; his idea of the best means of promoting family honor differs from his mother's. One is forced to conclude that his refusal stems from his childhood rebelliousness. He is not aware of this, of course. His guilt operates as a reaction formation, thereby distracting his attention from the real reason.

As a matter of fact, I believe an identical analysis can be made of his refusal to marry Chiyoko. Undoubtedly his uneven relationship with Chiyoko and her parents accounts for a large portion of his reluctance to propose to her, as even he too seems aware. This is at least the principal reason he offers when his mother asks for an explanation. "Chiyoko does not wish to marry me; and the Taguchis do not want her to, either." Once again, this decision seems to be the product of his relationship with his mother. When she tells him that Chiyoko has been promised to him since birth, naturally he asks why. "It's not for you," she replies tearfully; "I am asking you to marry her for my sake." He is unable, despite repeated questioning, to extract a clear-cut answer as to how the marriage will operate to his mother's advantage. No doubt he detects in her reticence a desire to conceal a secret, and it serves to reinforce his childhood premonitions. Making no further protest, he reconsiders the possibility of doing as asked. At the same time he considers various means of politely forestalling his mother should she take matters into her own hands and propose to Chiyoko in his stead. He knows full well that his mother has been far more disappointed by his reluctance to marry than by his refusal to work. The thought pains him, but he is convinced he has no other choice. Here we see the unconscious hostility he feels toward his mother who withholds the truth from him.

"Why Are You Jealous?"

Considering that Sunaga and Chiyoko were childhood playmates, one can expect that he would not view her in the same light as other members of the opposite sex; no wonder his mother's plans go awry. Although he knows that neither Chiyoko nor the Taguchis desire their marriage, he becomes quite upset on one occasion when Chiyoko announces in jest that her parents have found her a suitable mate. He considers the possibility that, unknown to himself, he has been in love with her, and she with him. This impression does not last long. He had always felt unequal to the purity and earnestness of her emotions, so he decides that, apart from the question of the wishes of either his mother or her parents, there is little likelihood of Chiyoko's and his coming together.

We can say, therefore, that at least outwardly Sunaga continues to act in a fairly disinterested manner toward his cousin.

During the summer of his junior year in college, however, he is thrown into considerable emotional turmoil when he visits the Taguchis at their summer resort in Kamakura and is introduced to Chiyoko's acquaintance, Takagi. Sunaga feels inferior in the presence of this refined young man who has returned recently from living abroad. He is shocked, moreover, to realize that his feelings of inferiority serve as a mask for his feelings of jealousy. Although this shock stems partly from the fact that he has never experienced jealousy before, it is far more frightening for him to realize that he does not possess the slightest desire to steal Chiyoko from Takagi. He feels that it would reflect poorly upon his character not to control his feelings, but jealousy rules the day and he is forced to spend two unpleasant days in the company of Chiyoko and her friend.

Sunaga spends the remainder of his stay and his return trip to Tokyo wrestling with the question of why he detests Takagi even though he has no desire to vie with him for Chiyoko. For the most part he holds to the proposition that "any woman willing to marry whoever proved victor in a duel for her hand was a woman not worth fighting for." How is he to explain his jealousy then? There is, he realizes, the matter of his pride. He secretly expects that Chiyoko will love him even though he does not reciprocate in kind. "It appeared at times," he says, "that she was the only person in the world who loved me." Takagi represents, then, a serious blow to his pride. He may not be a rival, but he threatens to become a menace to Sunaga's happiness. As Sunaga tells Keitarō, "I was about to be entrapped by a woman I had no intention of marrying. I felt my entrapment was inevitable as long as Takagi continued to appear in Chiyoko's company."

It is apparent from Sunaga's behavior on the day when Chiyoko escorts his mother back to Tokyo that he wants Chiyoko to rid him of this menace—or, in other words, to prove that she does not love Takagi. He is surprised to see that she would leave him behind, and he pays careful attention to the manner of her deportment vis-à-vis himself. Thus, when she explains that it is her responsibility as hostess to assure his mother's safe return, he complains in a manner that is almost coquettish (*amaete*) that he too had been a guest and should have been entitled to equal treatment. He further suggests that to have asked for equal treatment would have been to risk refusal. "If anyone was in danger of being refused it was I," Chiyoko says, looking to Sunaga's mother for confirmation. "What with the way you were acting so difficult, although you were an invited guest. You are really a little sick, you know." There is love implicit in Chiyoko's reprimand, and Sunaga even seems to derive pleasure from it.

During Chiyoko's stay at his house that night, he waits anxiously for her to mention Takagi's name in hopes that even the slightest reference will serve as an indication of her true feelings. She does not mention him, and Sunaga lies awake late into the night wondering why. Is she attempting to spare his feelings? Or does she have a different motivation? "Was she using Takagi as a decoy in order to ensnare me? Perhaps ensnarement was not her final goal and she merely enjoyed toying with my feelings. Perhaps she was telling me in her own way that she could love me if I became like Takagi. Or was she telling me how much she enjoyed the sight of two men fighting over her? Or, again, was she—by putting Takagi on display—trying to say that I should forget her once and for all?"

Finally Sunaga asks if Takagi is still staying in Kamakura when Chiyoko comes to announce her departure the next morning. "Does he bother you that much?" she says with a laugh. "You are a coward," she adds; and, when it becomes apparent that Sunaga does not understand why, she calls him a fool. His cowardice is not, she explains, the product of his introspectiveness; and she rejects the assertion that she despises him—"If anyone is contemptuous of anybody, surely it is you who despise me." When Sunaga denies ever having behaved toward her in a morally irresponsible or cowardly fashion, she is forced to explain herself further and, bursting into tears, she says, "You laugh at me because you think I am too forward for a woman. You do not love me. When it comes to marriage . . ." Sunaga attempts to interrupt, but Chiyoko insists on finishing. "Listen to me please. Perhaps you think that I am as much to blame as you are, and you may think that way if you like. I have no intention of asking you to marry me. But why, when you do not love me and have no intention of marrying me"—and here she pauses momentarily—"why are you jealous?" With that she begins to sob.

This scene might have proven highly embarrassing for Sunaga had Chiyoko been able to maintain her usual air of cool superiority. He might have turned several shades of red, or even prostrated himself at her feet in an attempt to apologize, but that would have been the end of the matter. Heretofore he had idealized her as "a woman who knew no fear" and he "interpreted her strength of character as solely and simply the accretion of her straightforward and gentle femininity." But the angry and tearful woman who sits before him seems to have lost not only her strength but also her gentleness. He is too weak a person to be able to understand weakness in another. There no longer seems to be any love left in Chiyoko's reprimands, and he begins to feel that he is hated and despised.

I believe that this scene provides ample material to explain the nature of his jealousy. His jealousy is, as he himself realized, related to the pride he derives from knowing that Chiyoko loves him. I believe, moreover, that his statement to the effect that he has no intention of vying with Takagi for possession of Chiyoko is an equally valid expression of his feelings. It is valid because in his unconscious mind he had already decided that she belonged to him. This unconscious sense of ownership explains why Takagi represents more of a *usurper* than a *rival* and why Sunaga perceives him as a menace to his happiness. Thus, despite Sunaga's repeated refusals to be drawn into rivalry, he daydreams in one episode of smashing a paperweight into Takagi's skull.

How is it that he comes to feel as though Chiyoko belongs to him? His words in the following passage are highly suggestive. "I do not hesitate to say it: no one but Chiyoko could make me feel my limitations so acutely. I am at a loss to explain, however, what aspect of her character reduced me to such straits. Perhaps it was her kindness." His remark is very much to the point. Sunaga enjoys the feeling that he can depend upon Chiyoko to treat him kindly (*amaeru*), although he does not love her in return. It is very likely that this tendency is related to the fact that he is an adopted child. Although he uses his prerogative as an only son to take advantage of his mother's love (*amaeru*), there remains an uncertainty and distance between mother and son that inhibits discussion of the true nature of their relationship. No such inhibition exists with Chiyoko and he is able to enjoy her favor in a straightforward and simplistic way. In a sense Chiyoko is more motherly than his mother. While ultimately the two cousins may not be blood relations, they share a far more profound kinship. As a result, Sunaga experiences great heartache at the prospect that Chiyoko is about to marry and leave him forever.

To Know the Truth

It becomes impossible for Sunaga to persist in his idealization of Chiyoko after their tragic confrontation. He had believed that she entertained only the purest good will toward him even though she was not interested in him as a possible husband. The confrontation produces a sharp break in their relations and Sunaga is forced to abandon his naive belief in her kindness—a step that is especially painful when we consider the extent to which he looked to her as a type of mother figure. He had always tended to introspection and vacillation, but now these tendencies assume abnormal proportions. I would say that he has reached the point where he requires psychiatric help.

Matsumoto reports one incident that supports this view. He notices that his nephew's coloring is extremely poor when he comes to call one afternoon. When at last he finds a convenient stopping point in his work, he searches about the house and finds Sunaga seated at a desk in the children's room. The boy is staring at the photograph of a beautiful woman in a ladies' magazine. The woman, he says, has kept him company for the last ten minutes; her beauty has the power to distract his mind from his personal problems. Matsumoto notes with surprise, however, that his nephew has yet to read the name printed at the bottom of the page. In other words, the picture could very well have been a fine oil portrait that one might admire. Photographs are, however, reproductions of real objects and, in the case of magazines, they are frequently advertisements. The names of the subjects are usually included, but "he was looking at the picture as a mere picture." Such behavior may not at first appear abnormal. It does reflect, as even a layman like Matsumoto quickly perceives, the alarming signs of indifference toward reality that are characteristic of mental illness.

These abnormal tendencies are most clearly observable in the conversation that ensues between Sunaga and his uncle several months prior to Sunaga's graduation from college. Matsumoto calls the boy to his house after his sister asks him to make one last effort to persuade her son to marry Chiyoko. Matsumoto, making no progress whatsoever, finally poses a hypothetical situation in which both the Taguchis and Chiyoko consent to the marriage. There is a brief silence during which Sunaga happens to notice the unusual look of utter despair that has crept across his uncle's face. Suddenly he asks, "Why does everyone hate me?" Matsumoto chides his nephew for talking nonsense, but Sunaga insists he is serious. "I am asking you because it is a fact . . . Am I not hated by you?"

It may be that Sunaga interpreted his uncle's momentary silence as an expression of disapproval and dislike—or at least this is Matsumoto's analysis of the situation. We would probably be more correct to imagine that Sunaga's mind is operating along slightly different lines. Since the previous summer he had become convinced that Chiyoko hated him. He could not believe that anyone could seriously suggest that she might still wish to marry him. Normally, he should have been able to dismiss Matsumoto's hypothetical offer, but the blow that he had sustained in his clash with Chiyoko has been of such severity and importance as to convince him that he is disliked by not only Chiyoko but everyone else as well. Matsumoto becomes, then, a ready target for his resentment.

Matsumoto rejects his nephew's charge because he believes it is based on a misunderstanding of his feelings. "Why should I hate you? You have known me since you were a child. Enough of this foolishness." Sunaga continues to listen quietly. "I am your uncle. Whoever heard of an uncle who hated his nephew?" With this, an expression of deep and utter contempt passes across Sunaga's face: Matsumoto has chosen the wrong words. He revises them, making a more general statement about the nature of avuncular love. "You are a well educated young man. You have a good head on your shoulders. But for some reason, you have a warped streak. You are to blame for it, and it needs correcting. It is unpleasant to see you behave this way." "That is what I said," counters Sunaga; "Everybody hates me, including you." At a loss for words, Matsumoto repeats himself: "There'd be no problem if you would rid yourself of that warped streak." But when it comes to what aspect of Sunaga's personality requires changing, once again he is unable to explain. "That is your problem. It would not hurt you to solve this one by yourself." Sunaga greets these words with a sharp reply: "How unkind you are!"

"I have considered the possibility that the trouble may lie with me," he adds. "I have had to do my own thinking since no one was kind enough to tell me the truth. I have thought about it day and night. I have thought about it so long that it has ruined my health and my sanity. Still I do not know the answer, and that is why I asked you. You say you are my uncle. And that you are kind to me. But what you have said to me now is far crueler than anything a stranger could ever say." Tears stream down his cheeks. "Very well, I have a warped streak. I knew that without being told or admonished. But I would like to know how I got this way. Everybody—you, Mother, the Taguchis—everybody but me knows and will not tell. I asked you because I trusted you above the others. But you have cruelly refused me. I will curse you for the remainder of my life."

Matsumoto decides then and there to reveal the secret of his nephew's parentage. He realizes that Sunaga is trying desperately, despite his tremendous fears, to grasp the truth, and now more than ever he deserves to know it. In revealing this piece of information, Matsumoto takes great care not to arouse Sunaga's enmity against his foster mother. His concern is to prove unnecessary: his story sweeps aside the air of uncertainty that had surrounded Sunaga's relationship with his mother and opens the way to freer communication. "I feel relieved and reassured now that I have listened to you and the mystery has been resolved. I need not be afraid or anxious again . . . At the same time, suddenly I feel forlorn and alone. I feel as though I am facing the world entirely on my own." One imagines that this newly awakened sense of isolation will operate as a major plus factor in terms of promoting Sunaga's future emotional growth. Once he achieves an awareness of himself as an individual, he should cease to feel the need to rebel against his mother and to lament the loss of Chiyoko as an object for his dependency wishes (*amae*). "No doubt I will cry when I return home and see my mother," he says. "The thought of these tears makes me unbearably sad." One imagines that these tears will reflect not sentimentality but his new awareness of his foster mother's love and his nascent sense of gratitude for it.

As we have seen, Matsumoto's revelation of Sunaga's secret has a highly therapeutic effect. He himself says, "It was a beautiful experience for me. That two people could totally unburden themselves is an event that shines gem-like against the backdrop of my drab past. I also imagine that it was the first time in Ichizō's life that he knew genuine consolation. There was a warm feeling in my heart after he returned home. I felt as though I had performed a charitable act." One feels that Matsumoto's story provides a highly valuable moral—that the truth has the power not only to console but also to be therapeutic. Merely reprimanding a person for a "warped streak" will not help to eliminate it. It is only through removing the falsehood that is built into a relationship and replacing it with the truth that trust will result and distorted thinking be corrected. This moral is particularly valid when it comes to the question of resolving conflicts between the generations. Young people cannot be expected to live in a responsible, meaningful way when the older generation denies them the truth about life. In addition, this moral explains why psychoanalysis proves effective. In its original form, psychoanalytic therapy consisted of making the unconscious conscious. Later it was realized that this formula alone was insufficient and various other theories were advanced. It is not possible to introduce them here, but suffice it to say that time has shown that they too have been found wanting. If anything, the original formula seems to have come closer to capturing in a few words the therapeutic significance of the psychoanalytic method. The process of making the unconscious conscious is the process of teaching a patient his own truth. To put that another way, the psychoanalytic method is therapeutic to the extent to which it succeeds in discovering the truth. Indeed, the truth has the power to set men free.

Chapter Seven

KŌJIN (The Wayfarer)

1912–13

Synopsis

The Nagano family lives in Banchō, Tokyo. Mr. Nagano is a retired government official, and he and his wife, Otsuna, have entrusted control of the household to their elder son, Ichirō, a university professor. Ichirō is married to Nao and they have a daughter, Yoshie. Jirō and Oshige, a younger brother and sister, also live at home.

A brilliant scholar, Ichirō is also a sensitive, difficult person. He seldom emerges from his upstairs study, and at dinner he is by turns taciturn and cantankerous. While Mrs. Nagano tries to humor her son, and Mr. Nagano to smooth over matters for the others, the family looks to Nao to find a means of calming Ichirō's nerves. She is unable or unwilling to do this.

As the novel opens, younger brother Jirō has been dispatched to Osaka to arrange a marriage for a domestic named Osada. He has little idea of what makes for a good match, and his thoughts are preoccupied with his friend Misawa, who has arrived in Osaka only to be hospitalized with ulcers. Jirō endeavors to be solicitous, but it soon becomes irritatingly apparent that Misawa is prolonging his illness in order to remain in the company of a young geisha in an adjoining sickroom.

No sooner has Misawa recuperated than Ichirō, Nao, and Mrs. Nagano arrive in the city to inspect Jirō's choice of a husband for Osada. Business completed, the family retreats to Wakanoura, a seaside resort on the Ise Peninsula.

Ichirō is irritable as ever and, taking Jirō aside, he complains that the source of his unhappiness is Nao's indifference. If only he could read minds, he laments. He suspects that Nao is secretly in love with Jirō and proposes a test. Jirō is to take Nao to a nearby resort to spend the night. When Jirō refuses, Ichirō presses for a display of fraternal loyality.

The following day Jirō and Nao set out for lunch in Wakayama under threatening skies. They have always enjoyed an easy rapport but, while Nao looks forward to this rare moment away from the family, Jirō is unusually nervous. At the restaurant he broaches the subject of Nao's relationship with Ichirō. She cries and insists that she has done her best. Jirō is moved to sympathy but, when a fierce storm arises and they are unable to return to Wakanoura, he feels uncomfortable at their having to spend the night together in a hotel.

The storm mounts in intensity. The lights fail, and candles are brought into their room. Nao clearly enjoys the adventure. If she must die, she says, better by a tidal wave or a bolt of lightning. "But most men," she adds retiring for the night, "are cowards in difficult situations." Unable to sleep, Jirō sits up chain-smoking. He reports to Ichirō the next morning that there is no question of Nao's integrity. As for a full account of the excursion, it will have to wait until after the family returns to Tokyo.

Several weeks later Mr. Nagano is entertaining company and boasts of his tact in dealing with others' affairs. He was asked to call on a woman whom a friend casually seduced and abandoned twenty years earlier. The woman was now blind, and the man wished to present her with money. Although the woman refused the gift, she desperately wished to know the reason for her rejection so long ago. Mr. Nagano assured her that, had circumstances been otherwise, the man would have married her.

The story meets the approval of all but Ichirō who later corners Jirō in his study and cites it as proof of their father's lack of sincerity. Jirō possesses the same trait, he says. Just as Mr. Nagano concealed the truth from the blind woman, Jirō has yet to report on the night at Wakayama. The brothers quarrel, and Jirō decides to move out of the house. Ichirō greets the news calmly and, making Jirō listen to a passage he reads aloud from Dante's account of the illicit lovers Francesca da Rimini and her brother-in-law Paolo, he states his conviction that Jirō has betrayed him.

Jirō goes to work for Mr. B's firm. He avoids the house in Banchō throughout the winter, but reports reach him of Ichirō's peculiar behavior at school. When Nao appears suddenly at his lodgings one spring night, he wonders if Ichirō has been violent with her. Finally, his parents call him home, and from Oshige he learns that Ichirō has commandered her to stand outside of his study door to participate in "telepathy experiments."

The family decides to ask H, Ichirō's best friend, to take Ichirō on a trip. Jirō asks H to record his impressions of Ichirō's behavior, and it is H's letter that constitutes the last chapters of the novel.

According to the letter, the two men have traveled around Izu Peninsula and are staying presently at a friend's summer cottage in Beniyagatsu. This is, after ten restless days, the first place that Ichirō is able to sleep. H realizes now that Ichirō is seriously disturbed. He believes that the cause of Ichirō's suffering is his heightened aesthetic, ethical, and intellectual sensitivity which renders him incapable of a more pedestrian attitude toward life. He feels that only religion offers hope, an alternative Ichirō rejects categorically. Unable to believe, and incapable of suicide, Ichirō suggests that his only choice is to sink into insanity.

The Insane Mind

In the mind of the layman the term insanity is commonly associated with the loss of both reason and a sense of propriety. The behavior of the insane is understood to be meaningless, and any attempt to accept it at face value would be considered a bit of lunacy. The psychiatrist who makes the treatment of the insane his speciality is also likely to define the term along similar lines. He would say that insanity represents a special type of mental state caused by a certain type of illness. Though the nature of the illness may not be apparent, he is inclined to believe that it is based upon physiological change and that, if the physiological imbalance can be corrected, the mental imbalance should also disappear. This is one conclusion resulting from many years of psychiatric observation, and its worth is not to be dismissed lightly.

But, while it is true that the health of the mind and body are intimately related and that physiological changes affect the psyche greatly, is this to say that psychosis is merely a passing symptom of physical illness? And that it has significance only as a disease in the traditional medical sense? Perhaps insanity has a logic of its own. This is one question, I believe, this novel is asking.

The question is first asked by the character Misawa. At some time in the past his family had the daughter of a friend living with them after the girl's marriage had ended unhappily. The girl was "withdrawn and melancholic" except, oddly enough, when Misawa was about to leave the house. She would follow him to the door and say, "Please, please come home early for me." She would repeat this request until Misawa told her to wait patiently for his return. This was extremely embarrassing to say in front of his family, but he took pity on the girl and tried to return as early as possible. "I made a point to approach her and say 'I'm home.'" As time passed, the family realized that the girl was mentally disturbed. "She had a very fair complexion and was very beautiful. Her eyebrows were dark black, and her eyes big and round and always wet with a rapt, faraway look. A sense of helplessness and sorrow seemed to float about them." They seemed "to cling to him as though to say, 'I am terribly lonely. Please help me.'"

When Nagano Jirō, the narrator of the novel, hears this story he asks if the girl behaved as she did out of love. Misawa replies that, while it was extremely difficult in her case to know what was sickness and what was love, he would like to believe that she did love him. He launches into the following tale of the circumstances surrounding her breakdown. "It's not clear whether her husband was given to dissipation or merely excessive socializing, but in any case shortly after the marriage he began to leave the house and stay out late. This seems to have been a source of tremendous mental suffering for the girl, but she tolerated it without a word of complaint. They say that because her mind was affected she would say to me in her illness what she had intended to say to him." Misawa does not want to accept this interpretation, however; for, as Jirō surmises, he had grown very fond of the girl. "That's right," Misawa says, "the more disturbed she became the more I grew to like her."

Several days later, when the Nagano family is traveling in the Osaka area, Jirō mentions Misawa's story to Ichirō, his older brother, in an effort to draw him into conversation. Ichirō is in one of his usual bad moods. Not only has Ichirō heard the story from his friend H, he also knows a fact missing from Misawa's version—namely, that Misawa had kissed the cold brow of the girl's head when she died. Letting curiosity get the better of him, Jirō presses for details. Ichirō knows nothing more and, when asked why he has kept the story secret all this time, he says summarily, "There was no need to tell you." Later that evening, as the two brothers are climbing into the bath, it is Ichirō who returns to the subject of the psychotic girl. He raises Misawa's question: Did the girl really care for Misawa? Or was she in her insanity saying to him what she could not say to her husband?

Jirō sees no reason why he should know but, as for Ichirō, he cannot but think that the girl was in love with Misawa. "That is," he says, "the way I see things," and offers the following explanation. "There are any number of things that ordinarily people would like to say were it not for etiquette or obligation . . . but if you were insane, think how much easier life would be . . . Let us suppose that the girl is the victim of insanity. Her usual sense of responsibility would disappear. In that event, she could say anything that came into her mind regardless of the consequences. That means what she said to Misawa was far purer and more sincere than the empty amenities we normally use." Jirō remarks that his brother's impressive logic is "most interesting." Ichirō demands to know, however, "whether, in fact, you think I am right or not. Hell, Jirō, this isn't an idea off the top of my head that is just interesting!" He says, "A man will have to drive a woman mad before he can determine what she is thinking," and heaves a painful sigh.

It is the thought of his own wife, Nao, that prompts the sigh. Ichirō suspects that she loves Jirō. He lacks proof, however; and that is all the more reason for finding a way to confirm his suspicions. On the next day, in fact, he confides in Jirō and, on the following one, asks him to test Nao's fidelity. The sad irony of this is that as Ichirō prepares to sacrifice his wife's sanity he himself takes the first step toward madness. We can imagine that when he speaks of the girl he is also speaking of himself. No doubt he would refuse to believe this, but the vehemence with which he castigates Jirō's lack of interest in his opinions speaks of the degree to which he identifies with his own theory. Certainly it is not academic interest that attracts him to the subject. He discusses it as though he were speaking on the basis of his own experiences.

Both Misawa's response to the girl and Ichirō's interpretation of that response strongly suggest that insanity is neither as meaningless and incomprehensible as is normally supposed nor is it a mere side-effect of physical pathology. Instead it possesses a meaning and logic of its own. While the insane may lose their minds, they are not without feelings. They may not ask for help in so many words but, as Misawa perceives, their seemingly mysterious behavior cries out for it; and, as Ichirō surmises, they may for a variety of reasons be giving vent to feelings that have built up inside. This is, interestingly enough, what psychoanalysis has to say about psychosis. What is even more fascinating is that nowhere can we find evidence that in writing *Kōjin* Sōseki was influenced by thought in psychoanalytic circles. One must conclude that his understanding of insanity stems from the same penetrating self-analysis that we find Ichirō practicing.

I never cease to be amazed at the fact that this novel discusses the deep implications of the phenomenon of transference at a time when psychoanalysis had yet to develop the concept fully. In other words, the interpretation that the psychotic girl says to one person what she cannot say to another constitutes a description of transference. It is true that Misawa rejects this interpretation because, one suspects, he is reluctant to tamper with the legitimacy of the girl's feelings. Actually, the unrealistic aspects of transference are often overemphasized because many people who think along psychoanalytic lines tend to define the term as the redirection of feelings for an infantile object to a present one. But, as Misawa perceives, any interpretation of a person's feelings as transference that fails to recognize the legitimacy of those feelings is tantamount to insult.

The irony of Misawa's rejection of the logic of transference lies, incidentally, in the fact that he has experienced the phenomenon firsthand. He is attracted to a certain geisha when he spends the first night of his arrival in Osaka at a party in the pleasure district. Several days later he has an attack of stomach ulcers. No sooner is he in the hospital than the same geisha—as though in pursuit—is also admitted with ulcers. Remembering how he had forced her to drink, he blames himself for her worsened condition. He visits her on the day Jirō comes to take him home and presents her with a handsome sum of money. In what is almost a soliloquy on his reason for displaying such largess to a stranger, he tells Jirō the story of the psychotic girl. Finally, he cries, "'I almost forget the most important point of the story! That girl—the geisha—she looks very much like the disturbed girl.' A smile that seemed to say, 'Now you understand, don't you, Jirō?' passed across his face."

Ichirō's Scheme

Ichirō has his reasons for believing his wife loves his brother. In the first place, Jirō and Nao had been friends prior to the wedding and an easy informality has existed between them ever since. Meanwhile, the relationship between Nao and himself continues to be shaky despite the birth of a child. Perhaps they were never well-matched, but living under the same roof with parents and in-laws does little to contribute to the success of their marriage. Jirō's presence, in particular, probably makes Ichirō unhappy. Still he does not complain. He is a product of the Meiji era and, as the eldest, undoubtedly he considers it natural that his family live with him. It does upset his sensitive nerves, however, to see Jirō acting conspicuously more relaxed with Nao.

We could not call Ichirō's behavior abnormal had he been merely jealous. It appears, in fact, that the relationship between Jirō and Nao attracts the attention of the entire family and once, when Jirō speaks sharply to his younger sister about getting married, she replies sarcastically, "Isn't it *you* who ought to be finding someone nice like Nao?" Later when he decides to move out of the house, Misawa scolds him for having waited so long—"It was wrong of you to have stayed around Nao and the others until now." Viewed in this light it seems that Jirō's behavior is not entirely above reproach. But what strikes one as most peculiar is the way in which Ichirō decides to reveal his pent-up jealousy. Rather than confronting Jirō with his anger and indignation, he asks him in a manner that is both extremely detached and full of confidence in his brother's character if "Nao is in love with you?" Jirō hardly knows what to say. He announces in a reproving tone that people must trust one another because, "when it comes to the question of what people truly feel, one can never know. No matter how great a scholar you are." Try as he may, Ichirō cannot believe in others. "I can only think, think, think. Jirō, please help me to believe." Ichirō's pleading strikes Jirō like that of "an eighteen-year-old."

At this point Ichirō mentions a favor. He will not elaborate until the following day when the two men take a walk. Prefacing his request with remarks about how difficult it is to ask and how Jirō must not allow himself to be shocked, he finally asks his brother to take Nao to Wakayama for the night to test her fidelity. When Jirō, greatly startled, refuses—"anything but that"—Ichirō says threateningly, "Very well then, but do not forget that I will suspect you for the rest of my life." Knowing that this leaves his brother with no alternative, he presses his request a second time. "Jirō, I trust you, but it's Nao whom I suspect. Unfortunately the one I suspect her of being in love with is you. I say unfortunately, but what may be your misfortune could be a stroke of luck for me. I'm fortunate in that I can trust you, otherwise I wouldn't ask. Don't you see the logic in what I am saying?" As Jirō listens he begins to suspect a deeper, hidden meaning. He suspects that, Ichirō's statement to the contrary, he is not trusted, and he refuses to undertake the test on the grounds that it would compromise his sense of honor. Ichirō insists that honor is irrelevant. "I'm not asking you to *do* anything." Jirō remains unmoved, but he does promise to speak to his sister-in-law. Thus it is arranged for the two to take an excursion to Wakayama the next day. That evening a storm arises and they are forced, as Ichirō wished, to spend the night apart from the rest of the family.

To ask a brother whom one suspects of overfamiliarity with one's wife to spend the night with her defies common sense. "There were moments," Jirō says, "when I decided that my brother was truly insane." It goes without saying that his amateur diagnosis eventually proves correct, but what we need to analyze is why Ichirō behaves this way in the first place. Taken literally, his reasoning is not, as he says, without logic. He trusts the brother but not the wife; if the brother can prove the wife's fidelity then he, the husband, can also trust her. At the same time, it is not without certain problems. First, why resort to this particular method? As I stated previously, there is nothing especially untoward about a little jealousy over Jirō. But then to ask him to test Nao's fidelity is quite extraordinary. Secondly, does he really believe that his brother will take his side in such a test? On a conscious level at least, Ichirō expresses complete confidence in Jirō and this is what prompts him to reveal his anxieties and ask for help. But his confidence is riddled with ambivalence and carries the seeds of distrust. We need only recall how he doubts Jirō when he blushes at the mention of Nao's name and how he threatens to distrust him as long as his favor goes unfulfilled. Finally, will the results of this test be convincing? Later developments suggest that they will not.

For when Jirō reports that "Nao's deportment is above reproach," suddenly Ichirō turns pale. This is a strange turn of events indeed if we are to believe what Ichirō had said earlier about his confidence in his brother. Although he ought to have looked relieved, his face becomes ashen and the air in the room becomes so tense that Jirō expects "either blows or a tirade of abuse" as he starts to leave his seat. It is possible for us to understand this only if we assume that Ichirō expects to have his suspicions confirmed rather than denied. We can also imagine that Ichirō has detected a subtle change in Jirō's attitude since the overnight trip and that his confidence in his brother is dwindling rapidly. Nothing reprehensible had happened at Wakayama, and Jirō is feeling quite guiltless. He stands now in a position of superiority from which he can coolly survey his brother's mounting frustration. When called to make his report, he sits quietly smoking a cigarette waiting for Ichirō to initiate the conversation. He would say later, "Without realizing, I let my sister-in-law's attitude infect me. I deeply regret having taken an attitude which I can neither retract nor redress." In other words, while Jirō and Nao had no physical contact, they became greater spiritual allies and Jirō comes to view Ichirō from Nao's perspective. One as sensitive as Ichirō could perceive this change immediately. Now he not only rejects Jirō's testimony but also decides that his brother is working in collusion with Nao.

Ichirō decides to say nothing. He will wait until the family returns to Tokyo before receiving a full report. He never does ask for it, however; and Jirō lets the matter ride. He does not know the hurt he has caused his brother, and he continues to make occasional cutting remarks without the least compunction. In one instance he remarks that Ichirō does not know how to humor his children. Ichirō agrees that, as a result of having devoted his life to scholarly pursuits, he has not learned how to humor anyone, including his parents and his wife. Jirō says laughingly, "But isn't it nice that your fine lectures compensate for everything." Sensing that Ichirō is unamused, he suggests that his brother thinks too much and that, weather permitting, the two should go hiking the next Sunday. One hardly knows whether he means to reprimand or to encourage.

It happens that guests are invited to perform *Nō* chants with their father, and Ichirō, Nao, and Jirō are asked to attend. After the chanting, their father relates a story illustrating his tact in handling a friend's personal affairs. One day, a few weeks later, the two brothers are talking when suddenly Ichirō begins to criticize his father. There is something shallow and vain about him, he says, and Jirō has inherited the same trait. As proof, he points to Jirō's feigned forgetfulness about reporting on the night at Wakayama. Jirō is hard-pressed to defend himself. He had not thought the subject worth mentioning, but if it is a report that is required then Ichirō shall have one. "I guarantee, however, that you will not find the specter you are looking for." Ichirō says he will not pursue the subject, but Jirō adds a reprimand in what appears to be a defense of his sister-in-law. "I believe that would be best for you both. And father too. Be a better husband, and Nao will be a better wife." At this Ichirō begins to shout at the top of his voice. "You fool. Like Father you may excel at getting on in the world, but you haven't the makings of a gentleman. Why should I listen to what a frivolous bastard like you would say about Nao?"

After this incident Jirō decides to move to other lodgings. Ichirō greets the news with exceeding calm. Quoting lines from Dante's love story of Paolo and Francesca with considerable feeling, he remarks that, although the world remembers the names of the illicit wife and brother-in-law, it has forgotten the name of the husband. "That is why, Jirō, the man who sides with morality and triumphs momentarily is also the eternal loser. And why the man who loses momentarily for having followed the dictates of his passions also wins forever. I will not even be momentarily victorious, however. I am the eternal loser, and you intend to go on living as the present, future, and eternal victor, don't you?" Jirō says nothing. He cannot understand what has prompted this latest attack. If matters are this serious, why not divorce Nao? The most curious aspect of this scene is, however, Ichirō's lack of jealousy. Not once do his words indicate resentment toward his brother, and there is no trace of the anger he displayed on the day Jirō first reported on the test. Of course that anger was, strictly speaking, the product of his disappointment at his brother's lack of devotion. But how are we to explain the almost fatalistic attitude with which Ichirō accepts what he believes to be an illicit relationship between his brother and wife?

Setting this question aside for a moment, I should like to introduce a story that I believes throws light on this subject. It is the tale of "El Curioso Impertinente" from Cervantes' *Don Quixote.* There were two gentlemen, Anselmo and Lotario, who lived in Florence and whose friendship was proverbial. Lotario acted as go-between and arranged for Anselmo to marry a beautiful girl named Camila. Anselmo's happiness with his new wife was boundless, but the natural tendency for his friend to refrain from visiting the newlyweds came as an unexpected disappointment. Anselmo entreated Lotario to think of their new home as his own and to come and go as he always had. Lotario continued, nonetheless, to keep his visits to a minimum, and Anselmo remained disappointed. "Though I ought to be happy," Anselmo complained one day, "I am sad. That's because I cannot believe my Camila is truly virtuous. I must prove she will not yield to temptation. You, Lotario, will you play the part of the tempter?" Lotario tried his best to convince Anselmo to abandon such madness but, realizing the futility of his words, he finally consented. The result defied everyone's expectations. Lotario and Camila fell in love, and Anselmo, his friendship betrayed, killed himself in a fit of indignation.

While I am certain that Sōseki had read *Don Quixote,* I have no idea whether he was thinking of this particular tale when he wrote *Kōjin.* Whatever the case, the story is particularly helpful in elucidating the motivation behind Ichirō's scheme. Just as Anselmo's friendship for Lotario is clearly homosexual, there is, I believe, an element of Ichirō's feelings for Jirō that must be defined as homosexual.

Ichirō's Illness

I have already mentioned the composure with which Ichirō receives the news of Jirō's leaving. It is, of course, the calm before the storm, for by this time he is closer to the breaking point than anyone realizes. One day Jirō hears the distressing news that his brother has been acting strangely at school. H, a colleague of Ichirō, has told Misawa that, although Ichirō had been known for the clarity of his lectures, of late he has been given to inconsistencies. Questioned about them, moreover, he would try to explain himself repeatedly, then put his hand to his head and say, "There is something wrong with me, my mind . . ." and stand for the longest time staring out the window. The story troubles Jirō enough to make him sound out his mother's feelings. His visits to the house become more and more infrequent, however; and finally one day his father brings him home. There Jirō learns that Ichirō is absorbed in research on telepathy. Even his parents find his behavior mysterious. The family decides to ask H to take Ichirō on a trip. As the two men are about to depart, Jirō asks for a letter detailing his brother's conduct. He probably fears that he is to blame for Ichirō's mental upset, and he wants to know if Ichirō despises him.

The anxiously awaited letter arrives eleven days later. The opening paragraph suggests that H was not particularly pleased at having to write but he felt compelled in view of the gravity of Ichirō's condition. According to the letter, Ichirō began to show abnormal symptoms as early as the first night of the trip. H suggested a game of *go* before bed but Ichirō decided against playing after studying the board for a moment. "Never mind," he said; and "a strange expression flitted across his eyes." In a few minutes he announced his intention to play, but no sooner had the game begun than he quit again. He smiled sheepishly when H gave him a worried look. "Your brother said he was caught in the dilemma of not wanting to do anything–let alone play *go*–and feeling he had to do something. Even before the game began he foresaw that he would be overcome by the feeling that he ought not to be playing. Yet he could not stop himself. Helplessly, he faced the board. No sooner had the game started than he became impatient. In the end the black and white *go* stones began to look like fiends that deliberately aligned and separated, parted and met, in order to torture his mind. Ichirō said that he was on the verge of madly scrambling them together to drive away the fiends."

This state of mind can be thought to correspond to the symptoms of inappropriate feelings, grimace, and ambitendency seen in schizophrenia. Unlike the typical schizophrenic, however, Ichirō is able to articulate his inner feelings fairly adequately and to explain them rationally. "There is nothing more frustrating," he says, "than to know your means are unrelated to your ends." It was not possible for H genuinely to sympathize with Ichirō's frustration because, as he pointed out, it is the shared fate of all mankind. H imagined he was consoling his friend, but Ichirō replied that "human anxiety stems from the advance of science. Never once has science allowed us to pause in its ceaseless, forward motion–from foot to rickshaw to carriage to train to automobile to dirigible to airplane. No matter how far we go, it will not let us rest. I am frightened nearly to death by the thought of where it all may lead."

Admittedly Ichirō's anxiety is the anxiety of the whole race. But he cannot accept it passively as the nebulous, invincible destiny that H means when he says, "Change and flux are man's fate." Ichirō sees himself as actively involved in the angst of mankind, which is racing endlessly down the railway of life ignorant of its destination. One feels that he is quite correct in his critique of modern technological culture and that H has missed the point. Ichirō could only smile when H admitted that he too was frightened. "The fear you say you feel is the theoretical kind. Mine is different. It is a mortal dread. It lives and beats in my heart." Still uncomprehending, H repeated his earlier assertion: "If this is the fate of all, why should you alone be frightened?" "Because in my lifetime," Ichirō said, "I will experience the destiny that mankind will arrive at only after centuries. I will experience it not only in a lifetime but also in a decade or a year or, worse yet, in a month or a week. You may think me a liar but take any moment of my life—thirty minutes or an hour—and examine it. It is the same hellish fate. All the anxiety of the race converges on me and I experience the dread of it every second on my life."

Certainly as psychiatrist Chitani Shichirō suggests in *Sōseki no Byōseki* (A pathograph of Natsume Sōseki),[15] Ichirō's statement includes expressions that can be appropriately labeled as anxiety tending to delusions of grandeur. I do not think, however, that such a diagnosis—and particularly Chitani's assertion that the anxiety is the product of depression—offers a full explanation of Ichirō's problem. Chitani's diagnosis would only confirm Ichirō's fear that his anxiety will be considered exaggerated and, perhaps, even a lie. Fortunately H did not question the legitimacy of Ichirō's feelings, as a psychiatrist would; and he concentrated on comforting his friend. "I gave up thought of all else. I wished I might save him from further anguish." But there is also little reason to expect that H would have known what to do. He felt unable to communicate with Ichirō and, quietly smoking his cigarette, he lost for the moment the desire to try. Ichirō shouted, "You are a better man than I."

H was more alarmed than ever. He felt "neither pleasure nor gratitude at these words of praise." As H finished his cigarette, Ichirō regained his composure and the two retired for the night. The next day Ichirō explained that H's face had "mirrored the natural, unadorned state of his soul" and that it helped him to escape his anxiety if only for a moment. To put that in other terms, we can say that he was delighted with the lack of contrivance on the part of his friend. H did not understand, for immediately he raised the subject of religion and suggested that the answer to Ichirō's problem lay in faith in God. This was clearly a contrived remark and, as one might expect, it aroused Ichirō's wrath.

Sensitivity to the slightest artifice in others is characteristic of schizophrenics, and frequently we find Ichirō behaving in such a manner. While walking in the hills, he alarmed H by pointing to lilies blooming in the underbrush and saying bizarrely, "Those are mine," or by pointing to the woods in the valley below them and announcing, "They are also mine." Finally, at the foot of the hill, he asked, "Where do our minds meet? Where do they part?" Taken by surprise, H was unable to express in words his feeling that Ichirō was searching for something that transcends human existence. In his confusion he muttered in German the proverb, "There is no bridge joining man to man." "That is what I expected you to say," said Ichirō. "He who is untruthful to himself can never be truthful to others . . . You deliberately set out on this trip as a babysitter, didn't you? I appreciate your good intentions, but hypocrisy disguised as sincerity will serve to drive a wedge between us." I am sure that H was genuinely shocked. Although he had never mentioned the role Ichirō's family played in planning the trip, Ichirō's sensitive nerves perceived that H's solicitude exceeded the demands of ordinary friendship.

Instead of taking Ichirō's advice and abandoning the role of babysitter, H seemed to feel compelled to try harder. There were times, moreover, when Ichirō did not resist. He related how lonely he was at home, and H tried to console him with an anecdote from the life of the French poet Stéphane Mallarmé. It seemed that a guest at a soirée inadvertently occupied the poet's favorite chair, and the loss of the seat upset the poet so greatly that he could not concentrate on his talk. This is a prime example of what occurs when a man like Ichirō permits his aesthetic, moral, and intellectual sensitivities to get the best of him; and H became more and more convinced that only religion could provide a means for Ichirō to escape his suffering. It happens that we also find Ichirō saying, "To die, to go insane, to embrace religion—these are my alternatives." Obviously he did not take the idea of religious salvation seriously because he said in the next breath, "I do not think I can possibly take up religion, and I am too attached to life to destroy myself. That leaves insanity. The future notwithstanding, am I sane now? I am terrified by the thought that I may already be insane."

It is generally held that psychotics do not recognize their own illness, and the assumption is generally correct. In Ichirō's case, however, he is acutely aware that something is wrong although he does not know what it is. "How fortunate was Mallarmé whose peace of mind was shattered by only the loss of a chair. I have lost nearly everything. Even my last possessions—these hands, these feet, this flesh—do not hesitate to betray me." These are surely the words of a psychotic. Ichirō requires a psychiatrist who will accept him as a sick person and who can understand what his disturbed mind is trying to say. H is not, of course, a psychiatrist and it seems he continually avoids the pressing question of insanity. It may be that he respects Ichirō too highly; or that, as a layman, he is afraid to deal with it. At any rate, he trotted out the subject of religion again and attempted to restore Ichirō's mental balance through belief in God. The overly eager manner in which he attempted to persuade Ichirō revealed the fragility of his convictions and earned him a slap across the face.

It was not the experience of being slapped that caused him to consider for the first time the possibility that Ichirō might be insane. Rather it was his awe at the consistency of Ichirō's logic which was infinitely more thoroughgoing than his own conventional thoughts on the subject of religion. "Beyond a shadow of a doubt I believe that your brother's brain is more neatly ordered than my own, but part of his personality has become disordered and the source of this disorder is the clockwork nature of his brain. I, for one, can admire the brain but question the mind. For Ichirō, however, the well-ordered brain *is* the disturbed mind. That is where I become confused. His brain is sound, but his mind is unstable. Dependable yet not to be depended upon. I wonder if you can accept this as a satisfactory reply. I too am at a loss to explain the situation to myself better." Thereafter H abandoned his efforts to reason with Ichirō and spent his time following him about. This included a wild spree of running and shouting in the driving wind and rain. Interestingly enough, Ichirō exclaimed repeatedly how "thrilling" it felt to get soaked to the skin. He also reintroduced the subject of religion, but whereas H's views were oriented toward the Judeo-Christian view of God, his leaned heavily on Indian Buddhist thought. He did not presume to represent himself as a believer, however. As the tears streamed down his face, he said, "Although I recognize the realm of the Absolute, it has continued to elude me no matter how successful I have been at defining my own world view. The only topography I have ever surveyed was in looking at maps, yet I have longed passionately to have the same experience as the man who snaps up his gaiters and traipses across the land. I know that I am stupid and inconsistent, but knowing that in no way prevents my blind struggling. I am a fool . . . You are a far greater man than I," he said in conclusion and prostrated himself before H.

The two men pursued their dialogue, and little by little H began to understand the feelings that underlay Ichirō's thinking. He wrote, "Could I but capture his mind so as to leave no room for a single inquiring thought. Capture it with visions of all the art of the world, the lofty mountains and great rivers of the earth, the beautiful women of the ages . . . When he says he desires to possess all things, doesn't he mean to be ultimately possessed by all? And that to possess absolutely is to be absolutely possessed? Only then will Ichirō, who cannot believe in God, find peace of mind." This is a truly brilliant insight. Even supposing that religion may have objective validity, it cannot solve Ichirō's problem after all. His problem precedes the religious question and relates to the nature of the psychology of the average human being. This is reflected in the envy he feels for ordinary people. He said, for example, that he envied Osada, a woman who until her marriage had lived and worked in his household as a maid. He rejected flatly H's suggestion, however, that happiness might lie in marrying a person like Osada because "marriage soon perverts a woman for her husband's sake." He confessed in a pained reflection that "as I say this I have no idea how much I have already ruined my own wife."

The novel's final scene suggests that a few days with an understanding friend produced a new sense of satisfaction. When Ichirō said, "Osada would be like you made into a woman," he hinted at a similarity between H and the girl. As H is a man and also in no danger of being corrupted, Ichirō was able to sleep for hours on end for the first time since the beginning of the trip. H used the time to write at length. The conclusion of his letter is highly significant. "I find it uncanny that Ichirō slept as I began this letter and that he sleeps as I finish it. I feel somehow that he would be fortunate never to awaken and, at the same time, that it would be sad indeed if he did not." H assumed the role of babysitter, and now at last Ichirō derives satisfaction from his friend's solicitude. H possesses Ichirō's heart at this point. But the question still needs to be asked: Will Ichirō remain happy this way? If not, what is to be done?

Humoring Others

While it is difficult to pinpoint the beginning of Ichirō's breakdown, we would probably be safe in choosing the time he feels betrayed by Jirō or even the day he proposes the test. It is more to the point perhaps to ask why Ichirō harbors extraordinary expectations of his brother. Otherwise, the novel might have ended differently.

The answer lies probably in an examination of the past relationship and home environment of the two brothers. We know that, as the eldest, Ichirō was raised by his old-fashioned father with all the rights and privileges befitting an elder son. His mother had also deferred to him because of his moody disposition. "She deliberately avoided hurting his feelings when correcting him for minor faults." On the other hand, she continued to treat Jirō like a child even after he was grown. "Although she scolded him fiercely —'How could anyone be so stupid?' she would say—she doted on him more." She gave him spending money without Ichirō's consent and retailored her husband's old clothes for him—"treatment that Ichirō found highly displeasing." When Ichirō was cross, "she would frown and whisper to Jirō that 'Elder brother's sickness has returned.'" Jirō enjoyed being made party to these little plots against his brother, but later he realized that Ichirō's anger "stemmed from more than a desire to be difficult . . . He had a sense of injustice as well . . . Still it was easier to operate through my mother. He always attached too many impossible conditions when I tried to obtain his permission by asking for it outright."

It is apparent that the father and, in particular, the mother have showered their natural affections upon the youngest while enshrining the eldest upon a pedestal. Undoubtedly this has made Ichirō jealous, but he has been adulated far too much to be able to indulge in an unsightly display of resentment. He cannot find an appropriate opportunity to release his anger, and any criticism of his mother's favoritism sounds self-righteous. Still he consciously endeavors to be friends with Jirō for, as he knows only too well, that is the way to maintain peace at home. He attaches great importance to their relationship despite the emotional barriers built into it. I think we can say that he attempts to identify with his brother. By tying himself to Jirō, he ties himself to the whole family.

The tragedy is that this triangular relationship involving Ichirō, Jirō, and their mother is transferred to and repeated in the relationship with Nao. Perhaps Jirō's previous acquaintance with Nao complicated matters, but the real problem is Ichirō's inability to be open and candid with his wife long after his younger brother has succeeded. In the triangular relationship with the mother Ichirō could perhaps forgive Jirō and settle for second best. But what about the relationship with his wife? Once again, Ichirō will not give vent to his jealousy much less his desire to drive his rival from the house. Jirō is the only one he can rely upon. It is Jirō—not Nao—to whom he confides and who must bear the burden of proving fidelity.

What does Ichirō hope to gain by this absurd request? Is it that he hopes to confirm his suspicion that he has been outmaneuvered again? But that is as good as proven. History has repeated itself. What occurred with the mother occurs with the wife. Regardless of what Jirō may say to the contrary, his attitude and behavior have convinced Ichirō of the validity of his suspicions. This is why, on the day Jirō reports his intention to move, Ichirō is able to state with exceeding calm, and without the slightest hint of irritation or jealousy, his belief that Jirō and Nao are allied against him eternally.

Why, then, the breakdown? It occurs because he feels he has lost his brother forever. He is sufficiently aware of the folly of his request, but he hopes that Jirō will understand his feelings and reward the trust he places in him. To overstate the case a bit, we might say he would have been satisfied with any reinforcement of his relationship with his brother at the expense of his wife. When this gamble fails, there is no choice left for him but to retreat into the realm of psychosis.

Readers may find it strange that Ichirō never discusses this important relationship involving his brother. He does not mention Jirō, for example, when in H's letter he complains of his isolation at home; only his wife and his parents are blamed. But does not this silence speak for itself? Has not the loss of the brother been an experience too bitter for words? Certainly Ichirō's silence on this point fits into the pattern of depression described in Chitani's analysis, but his breakdown is too severe to be classified as depression. His illness has gone beyond that stage and demonstrates a schizophrenic pattern.[16]

Except for his brother, he has no reason to remain within the circle of the living. As I mentioned before, when Jirō accuses him of being unable to humor (*ayasu*)[17] his children, Ichirō says, "It's not just the children. I do not know how to humor my father and mother, or even my wife." I find it extremely interesting that he applies this word to relations between adults because the verb *ayasu* is used ordinarily to describe an adult's way of loving or "cradling" a small child. Not only is he correct; he also shows considerable insight into the psychology of interpersonal relationships. Or, at least, interpersonal relationships in Japanese society. He gains this insight at considerable expense, namely, enshrinement upon a pedestal. As he never knew what it was to be spoiled or humored, he never learns to spoil and humor others; all of his youthful energies were spent on his studies. Only his tenuous relationship with his brother ties him to other human beings, and he believes it is a relationship based on a trust that transcends the manipulation involved in humoring and being humored. It is for precisely this reason that he loses faith in all human relations when he loses faith in Jirō.

What can be done to restore his faith? To humor and spoil him like a child at this late date will only incur his resentment. Yet, as H correctly observes, Ichirō requires someone who will possess him completely as in childhood. But then what? This is the problem that the author leaves not only for H but for ourselves as well to answer.

Chapter Eight

KOKORO (The Human Heart)

1914

Synopsis

This novel begins with an unidentified man's speaking in retrospect of his days as a student. His story goes back to a summer when, still in higher school, he is vacationing with a friend at the beaches in Kamakura. The friend is called home unexpectedly, and he is left at loose ends. One day his attention is drawn to a bespectacled, middle-aged man who appears on the beach in the company of a Westerner. He feels that he has seen him before and watches each day when the man returns for a swim. The man's glasses fall in the sand one afternoon and the boy uses the occasion to introduce himself; the next day, he takes the liberty of joining him in the water. This time the man initiates a conversation. The boy is jubilant. He addresses the man as "Sensei."[18]

He asks permission to continue their association back in Tokyo, but Sensei is not at home when he calls a month later. It is her husband's habit to visit a nearby cemetery this day of each month, says his wife, sending the boy after him. But, when the boy appears in the cemetery, a troubled look crosses Sensei's face. Asked who is buried there, Sensei says laconically, "A friend."

The enigma about this man attracts the boy and, during his years at the university, he becomes a frequent caller at Sensei's door. He learns that Sensei does not work and entertains no hope of having a family. An independent income makes it possible for him to live in modest comfort, and he and his beautiful wife, Shizu, lead a quiet, genteel life. But, whereas Sensei's wife seems to find fulfillment in making her husband happy, Sensei's reason for living is far less apparent.

In the boy's senior year, his father, a small landholder in the country, suffers a relapse of a kidney ailment. The boy returns home and does his best to amuse his father with games of chess, but he himself soon grows bored and his thoughts turn to Sensei—how provincial his father seems by comparison. When he mentions his friend, his parents assume that Sensei is a man of influence and send their son back to Tokyo toting a gift of mushrooms.

Sensei inquires after the father's health and, warning that the disease may be fatal, asks if the boy is assured of his rightful share of the family estate. He realizes this is an improper question, but he becomes excited and warns him to guard against his relatives. He had been cheated of his inheritance, and this is why he lives divorced from the world, distrusting humanity. In the past the boy had been surprised to hear this otherwise gentle man heap scorn upon his more successful peers; and once, when they saw lovers in a park, Sensei announced that "love is a sin." As to the causes of Sensei's misanthropy, he could only guess. This time he presses for an answer, but Sensei refuses to speak of his past.

The boy graduates from the university and returns home again. His parents derive great pleasure at the sight of his diploma, but a celebration party must be canceled when the Emperor becomes fatally ill. The father's health also worsens. Since the boy cannot return to Tokyo to look for work, his parents assuage themselves with the idea that Sensei will find him a position.

On the day of the imperial funeral, the nation is shocked to learn that the military hero General Maresuke Nogi has committed suicide in order to join the Emperor in death. The suicide strikes a chord in Sensei's heart too. He tells his wife that they have become anachronisms now that the Meiji era is over. Sending her off to tend to a sick aunt, he locks himself in his study and spends the next ten days writing a letter to his young friend.

The bulky envelope arrives at the boy's home. His father lies in a coma; the family keeps a bedside vigil. When at last the boy steals a glance at Sensei's letter, his eyes fall upon the passage—"I shall in all likelihood be dead by the time this reaches you." Overwhelmed with concern, he slips out of the house and rushes to the train station. As the train pulls out, he begins to read—

My parents died just before I went to Tokyo to enroll in

higher school. I entrusted my estate to an uncle who misused it. Outraged at this betrayal, I collected my assets and moved to the city permanently. I took lodgings in a house belonging to an army officer's widow and her pretty daughter and, when I recognized the palliative effect their company had upon my nerves, I insisted they take in a second lodger, a classmate named K who was also harassed with financial and family problems. Soon we both fell in love with the daughter. When it appeared that K would propose to her, I deliberately and brutally discouraged him in order to speak to the landlady first. Two days after he learned of my engagement to the daughter, he committed suicide. For the first time I perceived the enormity of my treachery—I was no better than the relatives whom I self-righteously despised. I kept my shame to myself and proceeded with the marriage. Unknown to herself, my wife has been a constant reminder of the wrong I did. Often I wished to die, but I confined the expiation of my guilt to the ritual of visiting K's grave each month. Now that the Meiji era has ended and I have in you a sincere confidant, I shall take my life.

The letter closes with a request that its contents be withheld from Sensei's wife.

The Youth and His Sensei

The initial encounter between the young student and the stranger he comes to consider his mentor, or "Sensei," occurs on a beach in Kamakura. It is the Westerner in the stranger's company that marks him out in the crowd. An interesting point—for even today we tend to think there is something culturally superior about Westerners and Japanese who associate with them. At any rate, the following day the youth finds the man alone and follows him into the water out of an impulse to make his acquaintance. The man takes no notice, and it is not until after two or three unsuccessful attempts on the boy's part that the two meet. The man's glasses happen to fall from under his robe as he is drying himself, and, in a flash, the youth scoops them up from under the changing room bench. At last he has his chance to speak to "Sensei."

During the next few days the two go swimming together, and

once the boy calls at his new friend's lodgings. The novel tells us that when he broaches the question of a possible previous encounter he is "strangely disappointed" to learn that his face is not at all familiar. "Haven't you mistaken me for someone else?" replies Sensei. Of course what the boy experiences is commonly known as *déjà vu.* In his case the experience is not particularly vivid, being nothing more than a vague longing for a friend; but it accounts for his disappointment and his request to be allowed to pursue the friendship back in Tokyo. Once again, Sensei's failure to appreciate the depth of the boy's feelings and to reply with more than a tentative "yes" comes as a blow to the boy's ego. The boy is not easily discouraged, however. "On the contrary. Everytime I felt anxious about our friendship I wanted to carry it further. I thought that sooner or later I would find in him what I was looking for." Back in Tokyo, he twice calls on his friend only to find him out on both occasions. On the second visit he asks for directions and sets out to meet him.

The youth becomes a regular guest at Sensei's household; each visit whets his appetite for the next. Of Sensei's ability to attract and captivate him, he says, "Even after we became friends, Sensei's manner towards me remained unchanged. He was always quiet. At times he verged on the forlorn. He had an air of impenetrable reserve, but I could not resist the impulse to become close to him." Many Japanese—particularly those of the pre-World War II and wartime generations—would consider this a typical and normal desire. They would say that young people often become infatuated with adults who appear to have the answers to life's questions and want to be in their company. They fail to realize, however, that this desire may not be universal. According to one foreigner who has lived and taught in this country since the early 1920s, Japanese students seem especially desirous of establishing personal relationships with their teachers. This may no longer be the case, and the change reflects perhaps the influence of a worldwide trend. Many students today greatly distrust their teachers and the older generation to which their teachers belong. They have lost not only an object to respect but also the desire to seek one.

Looked at through their eyes, the boy's feelings for Sensei probably seem childish and naive.

So far we have discussed the novel from the perspective of our own time. It may come as a real surprise therefore to discover that Sensei—a man of the Meiji era—likewise refuses to accept the boy's feelings as typical. Instead, he offers a brilliant interpretation of their psychological significance. He asks why the boy's visits are so frequent for, as he later says, he does not believe in his own worth and seeks the answer outside of himself and in the personality of the boy. "I am a lonely man . . . and I wonder if you might not be lonely too. Still, I am older and content to be less active. But what about you? You are young and want to move about to see what you can. In moving about, you imagine you'll discover something." Although the boy denies any suggestion of loneliness, Sensei is unconvinced. "Youth is the loneliest time of all. Otherwise why would you come here so often? . . . I suspect that you go away from me feeling lonely too. I do not possess the power to get at the roots of your loneliness. You will have to look elsewhere for the consolation you seek. Before long you will come here no more." Sensei smiles sadly as he says this.

Sensei repeats this idea on several occasions. When discussing the subject of romantic love, he suddenly announces that it is a sin to love. Asked why, he says, "You will understand soon enough, if not already. Your heart has been stirred by the vision of falling in love for some time now." The boy cannot fathom the object of this love, but Sensei reminds him of his visits. Admittedly Sensei fills an emotional void in his life, but surely his visits have nothing to do with love? "They are a step in your life toward love . . . As a step toward embracing the opposite sex, you come to me, a member of your own sex." The youth insists the two feelings are different, and, equally insistent, Sensei says they are the same. "As a man I am incapable of giving you the satisfaction you need. And due to special circumstances in my own life, I am even more inadequate. I am sincerely sorry. There is no way I can stop you from going elsewhere. In fact, I welcome it."

These admonitions have the effect of encouraging, rather than dampening, the boy's idolizing of his friend—"I respected his opinions more than those of my university teachers. He who quietly pursued his solitary path seemed far greater than the eminent professors who tried to give direction to my life from the podium" —and Sensei is forced to repeat his warnings. "You should be more objective in your evaluation of me . . . You are like a man with a fever; when the fever breaks your affection will turn to revulsion. Your opinion of me now is distressing enough, but I am even more pained at the thought of your inevitable disillusionment." This prompts the boy to ask if he appears insincere and unworthy of trust. "I cannot trust others," replies Sensei, "because I cannot trust myself. There is nothing I can do but curse myself." And in even stronger terms he warns, "Above all do not rely upon me too much. You will regret it in time. Let a man feel betrayed and he will seek his cruel revenge . . . The memory of having knelt before me will haunt you and drive you to trample me underfoot. Rather than face the insults of the future I would forego the admiration of the day. Better to be lonely now than utterly rejected later. This is the price we all must pay for being born in an age rich in freedom, independence, and egotism."

From a psychoanalytic point of view, this is an extremely interesting dialogue. Sensei assumes the role of the analyst; the boy, the analysand. Sensei sees the boy's infatuation as originating in feelings of loneliness and frustration, or a type of homosexual feeling. If we recall for a moment, moreover, how the youth had wanted to meet the stranger on the beach and had seized upon the opportunity to retrieve his glasses, we see that his actions are remarkably similar to those of a man who wishes to initiate a conversation with a woman who has taken his fancy. Of course, the boy rejects Sensei's analysis and makes no attempt to understand it; in fact, he probably never does come to fully understand Sensei's point. By the same token, Sensei never fully assumes the role of analyst because, in addition to offering advice, he also wants to make a confession. He is a lonely man. He cannot trust himself. There are special circumstances surrounding his case. With each mounting admonition his self-revelations become more dramatic and detailed. When he hears, for example, that the youth's father is seriously ill, he urges that the estate be settled without delay "because one never knows how people will behave when money is at stake." His tone is agitated, and he admits that he is easily upset by talk of estates and inheritances. "I shall never forget that I was cheated once by my own relatives. While my father lived they acted decently, but no sooner was he in the grave than they turned into thieving maggots. The humiliation I suffered as a youth is still with me and will be until the day I die. I shall never forget it. Still, I have not taken revenge. No, come to think of it, I have done something worse than that. I have come to despise not only my relatives but the whole human race as well. That is revenge enough for one man."

Once more Sensei's words seem to allude to something deeper —even philosophical—and the boy decides to ask about it. Sensei denies having anything to conceal. "It seems to me that you make a serious mistake in confusing ideas and opinions with past history. I may not be a rigorous thinker, but I do not hide the few ideas I have worked out over the years. I have no reason to. If you are hinting that I relate to you the details of my past—well, that is altogether different." It is interesting to note that, while the boy does not specifically request this information, Sensei feels as though he does. This reaction undoubtedly convinces the boy that the key to his friend's way of thinking lies in a knowledge of his past. Thereupon he asserts that ignorance is the source of his confusion and that Sensei's opinions are meaningless when divorced from their past. Sensei is dumbfounded: "You are a very audacious young man." The boy hastens to assure him of the sincerity of his desire to learn from others' experiences. "Even at the cost of raking up ancient history? . . . Are you genuinely sincere? I have come to distrust everyone because of my past experience. To be frank with you, I distrust you too. But, perhaps because you seem so straightforward, I do not want to distrust you. I would like to have one friend I can really trust before I die. Can you and will you be that friend? Are you genuinely sincere?"

At this point roles are reversed. It is no longer the boy, but

Sensei, who is the patient seeking help. As his own words later suggest—"A knowledge of my past may not be as valuable to you as you think. You may be better off not knowing"—talk of the past will stem first and foremost from his great longing to believe and trust in at least one other human being and only secondarily from any merit it may have for the listener. When he says, "I cannot tell you just yet. Don't expect me to tell you until the proper time," he acknowledges the presence of a psychological block that inhibits his secret desire. Only by laying his life on the line can he free himself.

Sensei vs. the Father

The boy returns home before the beginning of the winter vacation because of the worsening condition of his father's health. The father is better than expected, and after several days at home the boy grows bored. His mind turns to thoughts of Tokyo and Sensei. "Strange as it may sound, my heartbeat seemed to quicken and grow stronger, as though Sensei were able, through some delicate mechanism of the psyche, to send me encouragement . . . I thought of him in comparison with my father. Both men led quiet, peaceful lives. Indeed, they were so quiet that, for all the world cared, they might never have lived. They were, in society's terms, nothing at all." When it comes to the question of whose influence is greater, however, there is a decided difference. "My sick, chess-playing father could not keep me amused," whereas "it seemed no exaggeration to say that Sensei's strength had entered my body and coursed through my veins." He is startled, "like a man discovering a great truth," at the recollection that he is after all his father's son, and not Sensei's.

Tokyo has changed him and he no longer fits in at home. "It was like, as the old saying goes, a Christian in the house of a Confucianist." He tries for as long as possible to hide the fact that his interests no longer coincide with those of his parents, but finally, unable to bear the strain, he leaves for Tokyo. The following summer he graduates from school and returns home only to be plagued by the same feelings. The immense pride his father derives from the sight of his diploma strikes him as repulsive and boorish, and he resists his parents' proposal that they invite the neighbors to celebrate. "Don't go to a lot of trouble for my sake," he says. "If it's for me, I'm saying, don't bother. But if it's because you're nervous about what people will say, that of course is a completely different matter. Who am I to demand something that may hurt you?" His father smiles bitterly and comments that "learning makes a man argumentative." It is not until later that the boy realizes "how abrasive my words were. I thought it was my father who was being difficult." For better or for worse, the celebration has to be canceled when news arrives that the Emperor has been taken critically ill. In a fit of loneliness, the boy sends off a rash of postcards and letters. Sensei is among the recipients.

The father is frequently querulous. He knows that this son will, like his elder brother, move far away to find work, and he laments the lack of someone to assume control of the family house. "You know, sending one's children to school has its drawbacks. You take all the trouble to get them educated, and then they never come home again. You might almost say that education is the surest way of separating children from their parents." No sooner are the words out of his mouth than his son asks for a monthly allowance to defray the costs of searching for a position in Tokyo. "In my day," the father continues, "parents were supported by their children. Today, it's the children who are endlessly supported by their parents." This prompts yet another comparison in the boy's mind. "I knew almost all there was to know about my father. I would miss him if we were separated but that sadness would not be greater than any other devoted son's. By contrast, I knew almost nothing about Sensei . . . He stood in the shadows, and I would not be content until I brought him into the light." Impatiently, the boy prepares to leave. His father takes a turn for the worse on the day of his departure, and the trip has to be postponed. The father's condition worsens with each passing day. The other children are called home, and the family bides its time. A telegram arrives from Sensei asking his young friend to return to Tokyo. This is not possible, and the boy wires a negative answer. Several days later a large, heavy envelope arrives in the mail. When the boy realizes it is a posthumous letter from Sensei, he deserts his dying father and boards the next train to Tokyo.

As we have seen, the boy returns home to discover to his surprise the depth of his emotional involvement in his friend. We need only recall how uncannily accurate Sensei's pronouncements were. When the boy says that his blood is strangely quickened by Sensei or that his life courses through his veins, he is clearly "like a man in a fever," infatuated with his friend. Back at home and acutely aware of his inability to relate to his parents, he experiences the loneliness and spiritual want that Sensei had cited as the reason for his visits. How vexing this realization must have been now that he was placed in the unhappy situation of waiting for another to die. Moreover, the drama that Sensei had warned against was being acted out with his own father. As a child he had once played at his father's feet and worshiped him, but now he is repelled by his boorishness. It is for this reason that he can say matter-of-factly that he will not miss his father. It is also why he can abandon him in a whirl of concern for his Sensei.

The fact, incidentally, that he frequently associates and compares the two men suggests that, despite outward dissimilarities, the two are intricately bound together in his psyche. In other words, his feelings for Sensei constitute father transference. Feelings that had been directed toward the father and that had ended in disillusionment and frustration sought a new outlet and object in Sensei. The boy's *déjà vu* experience is, for example, a foreshadowing of this. Of course a layman like Sensei cannot be expected to have discussed the subject in these conceptual terms, but he is aware that the boy's feelings stem from a deep personal need. He believes, moreover, that the need borders on love, and that like all love affairs it will end in disenchantment. It is this belief that makes him so persistent in his admonitions.

Since the Meiji era it has become quite common for young people to grow up to reject their parents in favor of a new object of respect. This may be a worldwide phenomenon. But, because Westernization occurred so rapidly in this country, the phenomenon has been more exaggerated. One wonders how many parents have educated their children only to see them grow up with values lamentably different from their own; or how many young people have grown to secretly despise parents they once loved. The use by young people of such expressions as "the old man" (*oyaji*)[19] and "the old lady" (*ofukuro*)[20] point to this state of affairs. True, these expressions reflect a fondness arising from being flesh and blood, but they are devoid of respect. It is the schoolteacher with his new supply of knowledge who is respected. Or, when he is found wanting, it is a person like Sensei. Otherwise young people may, as is often the case in contemporary society, turn to revolutionary ideologies. As the example of modern Japan makes clear, the distance between the generations becomes significantly greater whenever a part of a nation's spiritual heritage is for one reason or another on the verge of collapse.

The problem is that, having lost faith in the older generation, the young still desire a new object, or idol, to respect; or, in Freudian terms, they seek a father figure. Sensei clearly understands this. He does not wish to be made an idol: idols are vain things and doomed to destruction. His admonitions are quite consistent in this respect, and it can be safely said that, of the many emotions that motivate him to write a final letter, one is his desire to make this point clear once and for all. He wants the boy to know how utterly disappointed he is with life and how he has come to hate himself and all humanity. Earlier he had said that he would forego "the admiration of the day" rather than face "the insults of the future," and his letter represents his final attempt to insure against being treated as an object worthy of respect. He is no longer capable of respecting or admiring any man or thing. He is a man done with idols.

There is, however, one flaw in his logic. At the point when supposedly he has become totally disillusioned with life and is about to draw the logical conclusion of his thinking and choose death, suddenly we hear him speak anew of respect. The old, forgotten words pour from his lips. In the opening passage of his letter he writes, "Though I did not scorn your opinions, I could not respect them either. Your ideas lacked depth; you were too young to have acquired much experience. Sometimes I laughed. And not infrequently, your own dissatisfaction showed in your face. Finally, you pressed me to unroll my history before you like a great scroll. *Then and there, for the first time,* I respected you. I was touched by your unabashed determination to reach out for some vital element in my soul. You wanted to tear open my heart and sip its blood. I was alive then. I did not want to die. I refused you and waited for the morrow. *Now, on my own, I am about to tear out my heart and let your face be splashed in its blood. I shall be well content if, when it has beat its last, new life dwells in you.*" [Doi's italics] The irony of the story is that the letter—and the disclosure of the darkest side of Sensei's personality—fails in any way to diminish the boy's respect for his mentor. Let us return to this curious contradiction later.

Sensei's Past

Sensei was the only child of well-to-do parents who died of typhoid fever shortly before he went to Tokyo to higher school. He had been raised in a liberal fashion, and the untimely death of his parents came as a shock. He records that at the time, "I was completely without knowledge and experience of the world. Furthermore, I lacked good sense." He did have the habit of trying to unravel things, of turning them over and over and studying them. The novel does not say at what age or for what reason he began to show these compulsive tendencies, but they were probably part of his inborn personality. His early home environment also probably encouraged their development. To be raised in a "liberal fashion" is really another way of saying that as a child Sensei was left to do as he pleased. Being an only child, he inevitably spent much time alone and this led to his thinking to himself a great deal. This was probably the origin of his compulsive probing into the nature of things.

After Sensei entered school in Tokyo, his paternal uncle moved into the family house and assumed charge of the estate. Sensei was happy with the arrangement since his parents had always considered his uncle reliable; there was no reason to suspect he might act otherwise. The first summer vacation at home was pleasant. There was one incident, however, that "cast a slight shadow over it." His aunt and uncle urged him to marry. In their opinion he should marry as soon as possible in order to inherit his father's estate. Marriage seemed like a problem of the distant future for a student who had only begun higher school, and Sensei returned to Tokyo without giving indication of consent. A year later, his uncle raised the question again. This time he had a bride in mind: his own daughter. He claimed that the arrangement had been the wish of his late brother. Despite his fondness for his cousin, Sensei found it difficult to think of her as his future wife. Besides, the modern climate of Tokyo had made him a convert to the notion of love-marriage. He rejected his uncle's proposal that they announce the engagement then and postpone the wedding until after graduation.

The third summer Sensei was totally unprepared to find that the marriage question had become a source of unpleasantness. He believed that he would be greeted with open arms. After several days, however, his uncle's attitude suddenly struck him as strange. When he thought about the matter, the whole family seemed to be acting peculiarly, and he wondered whose feelings—his or theirs—had changed. Or was it that his parents were trying to open his eyes from the other side of the grave? "I began to feel that unless I acted quickly all would be lost. I felt I owed it to my parents to learn what had happened to their estate." His uncle was always too busy to talk, and it seemed to Sensei that he was being avoided deliberately. He began to make inquiries, and his suspicions were heightened by rumors that the uncle kept a mistress and had failed in business. Sensei finally forced a confrontation which ended, as we might expect, in a severance of relations. The belief that he had been cheated became fixed in his mind, and he concluded that his cousin had been offered as bait. Leaving the remainder of his possessions to be disposed of by a friend, Sensei bid farewell to his home forever.

There is little point to our disputing the veracity of Sensei's account, but it does contain several inconsistencies. First, he does not clearly specify the manner in which he was cheated. Second, when he says it was in his financial interest to marry his cousin, does he mean that he would have profited from the receipt of her dowry? Or that he had never really been defrauded? In either event, the statement contradicts a second one in which he says that "talk of marriage stemmed from his uncle's blatant greed rather than any financial advantage that would accrue to either party." It seems entirely conceivable that, as the boy's guardian, the uncle might have favored the marriage for reasons other than economic. Marriages between cousins were not uncommon in that era. There is nothing wrong, of course, with the boy's rejecting the offer. But it does seem to be asking too much then to expect to be treated as cordially as ever. We already know from the case of the young university student what happens when a young man goes to Tokyo. Sensei, too, had breathed the freer air of the city and become a devotee of the love-marriage. Surely he should have known how his uncle would react. But he had been raised in a liberal fashion and did not know how to take others' feelings into account. Indeed, this was the source of his tragedy.

Logical as this may seem to us, it does not prove that the uncle was acting on purely altruistic grounds or that he was innocent of fraud. Nor does it prove that he spent his nephew's money on his mistress or his business losses, although as guardian he might have exercised considerable license in handling the estate. What can be said with a degree of certainty is that Sensei's reasons for doubting his uncle were irrational. His awareness of a change in attitude came "all of a sudden . . . without warning or time to prepare, totally unexpectedly." He never stopped to think that the change in his uncle was really a change in himself. And it is precisely because he refused to consider the question of his own behavior that the change struck him as so abrupt. In the early stages of schizophrenia patients often speak of sudden changes in their environment, and one wonders if Sensei was not experiencing the same phenomenon. In that event, his story would constitute a type of delusionary experience.

There are, moreover, two facts that support this interpretation. The first is the peculiar agitation he displays when he speaks to the boy of his past. In a scene that is almost out of character, he speaks movingly of these long forgotten events as though they happened only yesterday. The second is his state of mind at the time he left home. He writes, "I was a confirmed misanthrope when I left for the last time. The conviction that one could not rely upon others must have already been deeply implanted in my mind. I regarded my aunt and uncle and my other relatives as typical of the whole human race. I found myself eyeing the other passengers on the train suspiciously and, if one happened to speak to me, I doubled my guard. I felt weighed down, as though I had eaten lead. Yet my nerves were on edge . . . Even after I moved to my new lodgings, I found it difficult to quiet them. The way my eyes shifted about in their sockets became a source of embarrassment. Oddly, while my mind and eyes worked overtime, my tongue ceased to move at all. I sat taciturnly at my desk and watched the others like a cat. At times I felt sorry for them for being under such constant surveillance. I would tell myself that I was like a pickpocket who was too cowardly to steal."

As this quotation makes clear, Sensei suffered from what psychiatry once called ideas of reference of a persecutory nature. No doubt he would have viewed his behavior as resulting from the experience of having been cheated, but it is equally plausible that the defraudation was part and parcel of the same delusion of persecution. The abnormal symptoms gradually abated, however, in contact with the calm ways of his new landlady. "It came as a great stroke of good fortune that she and the rest of the household paid no attention to my deeply skeptical nature. I began to feel easier because there was nothing about the house that seemed to warrant suspicion." Unknown to herself the landlady had employed the best method of nursing a mentally ill person back to health. Of course her efforts proved only temporarily effective, for soon Sensei was besieged by a new set of anxieties: he had become attracted to her daughter. He was not content to enjoy this long-awaited first experience of romance; instead, he felt "strangely uneasy" and "nervous" in the daughter's company and he was confused and perplexed by the fact that the landlady seemed to be unduly encouraging the relationship at the same time she guarded against it. He attributed her behavior to the incomprehensible workings of the feminine mind—a conclusion that in no way answered the question of what he ought to do about his almost worshipful love for the daughter.

Later it occurred to him that a mother might feel both ways. Perhaps the landlady had always trusted him and he had been wrong to think ill of her. He allowed himself one day, despite considerable misgivings, to answer her questions about his family. Suddenly he was attacked by doubts that she too might be trying to foist her daughter on him. "She, who had always seemed so kindhearted, loomed in my mind as a cunning schemer." He began to suspect that the daughter was no less a schemer and that the two women were plotting behind his back. The thought of being fooled a second time left him almost at wit's end. He was, nonetheless, extremely jealous whenever men called at the house. Often, he made up his mind to ask boldly for the daughter's hand; as often he let the opportunity pass. It annoyed him to think he might fall into their trap.

In this highly emotional state, he seized upon the project of enlisting his hometown friend, K, to room with him. An adopted child, K had argued with his foster parents over plans for his education and, as a result, had been disinherited by his real parents as well. Sensei admired K's determination to live his own life, and it deeply distressed him that K's mental health was about to be undermined by economic hardships. He wanted to help his friend and, on the grounds that K was too proud to take money, he decided to have him move into his lodgings and live there free of charge. It is interesting to note that Sensei insisted on this plan in spite of his landlady's opposition. This suggests that he no longer considered himself a mere boarder and that he knew his request would not be denied. One can also imagine that the company of the two women had become an intolerable threat and that he unconsciously welcomed an ally of the same sex. K's move did, in fact, temporarily ward off Sensei's anxieties. The women were mobilized into taking care of their new lodger, and it was "with exceeding pleasure" that Sensei watched K relax and become more human.

Yet his peace of mind was short-lived. As K improved and became more friendly with the members of the house, Sensei's outlook grew grim: he had become jealous of K's friendship with the daughter. K was now a rival, not an ally. Sensei insisted that they take a trip to the beaches of the Boshū Peninsula in order to remove K from the house and reaffirm his friendship with him. But a breach between friends is not easily mended, and on the trip Sensei behaved like a man possessed. He shouted "like a savage," and once, on a high cliff, he grabbed K by the neck and seemed about to hurl him into the sea. K remained perfectly calm—a coolness that, as supposed proof of his confidence vis-à-vis the daughter, made Sensei more nervous. Sensei wanted to confide in his friend, but K's "strangely aloof attitude" kept him at bay. It is true that K considered Sensei "a fool" for refusing to share his devotion to spiritual training, but it also seems that Sensei hesitated to raise any subject that threatened to undermine their friendship. He later reminisces that "there is a kind of inertia built into a friendship that develops among friends pursuing similar academic interests, and I confess I lacked the courage to try to overcome it."

Even after the trip, Sensei became extremely jealous whenever K had an opportunity to be with the daughter. He gave renewed consideration to the idea of asking for permission to marry her only to abandon the plan out of fear that the girl was already in love with K. Then one day K told Sensei of his "heart-rending love" for the girl. The confession was, literally, a bolt from the blue. Incredible as it may seem, it appears that Sensei had never taken his friend's challenge seriously. He was taken aback completely, and no doubt he felt betrayed. He was no match for K, who now seemed like "a sort of fiend." Thus for several days he observed K's movements closely. He feared that K had already spoken to the landlady and only after he had ascertained otherwise did he dare to ask about his future plans. K's reply was vague and indefinite, and later when the two men went for a walk K asked for advice. Realizing that his friend was dejected and not feeling his usual confidence, Sensei repeated the words that once had been applied to him. "The man who does not endeavor to uplift himself spiritually is a fool."

In short, Sensei had his revenge. Yet it did not bring him peace of mind because he worried that K might decide to abandon his spiritual commitments in favor of marriage. Sensing there was no time to lose, he feigned illness in order to create an opportunity to speak privately with the landlady who, without much hesitation, agreed to let her daughter marry him. It was not until Sensei saw K, however, that he felt conscience-stricken. Despite a strong urge to beg forgiveness, he said nothing, and for the next several days he remained in low spirits. Then he learned that K had already heard the news from the landlady, who wished to know why his friend had not been told before. Two more days passed without a word between the two men. K seemed the same as ever, and Sensei could not help admiring his composure. He felt that, although he had won the daughter, he had lost "as a human being." He imagined how contemptible he must seem and blushed with shame. On the night that he finally resolved to speak to his friend, K committed suicide. There was a brief note explaining the reason for his death and asking Sensei to assume responsibility for the funeral arrangements. A great shudder passed through Sensei's body. "I felt almost the same sensation as the day K first told me he loved the landlady's daughter . . . I thought, 'It's too late!' The whole future course of my life flashed before my eyes illuminated by a cold, eerie light. There would be no appeal, no annulment." As the novel indicates, Sensei never does recover from the shock of K's death.

Sensei's Suicide

There is a very complex side to Sensei's feelings about K's death. Naturally he felt responsible; it was as though he had killed him himself. In particular, he felt conscience-stricken about the role his own deviousness had played. He had destroyed forever the image of himself as a "fine, outstanding human being," unique among men. It would be a mistake, however, to attribute his emotional reaction solely to feelings of guilt. I say this because his initial response was to be taken aback—to feel outwitted, outmaneuvered—as he had on the day K confessed his love for the daughter. In other words, despite the devious methods Sensei used to defeat his friend, it was K who proved the ultimate victor. It must have been with a terrible sense of irony and frustration that Sensei read the words, "I have chosen suicide because I am too infirm of purpose to live up to my ideals." For what did they imply about Sensei's ideals and his ability to live up to them? K's death loomed like a great question mark over his soul.

Shortly thereafter Sensei graduated from school and, in less than six months, he married the landlady's daughter. Yet he could not enjoy his new-found happiness. Instead of helping him to forget, his wife served as a constant reminder of their dead friend. Sensei began to devote more time to thinking alone, and, when his wife misinterpreted this as a sign of his displeasure, he was at a loss to explain. Often he stood on the verge of unburdening himself, but at the last moment he would relent. "I could not bear to sully her thoughts with the memory of a thing so ugly." While in one sense he was being considerate, in another he was tacitly preventing his wife from ever learning the true nature of his relationship with K. In fact, a cold chill ran down his spine at her suggestion that they visit K's grave together; and she never did accompany him on his monthly visits to the cemetery. It can be safely said that there was something irregular about the nature of Sensei's love for his wife. Before the marriage he had doubted the sincerity of her feelings for him, and he found it extremely difficult to decide to propose to her. It was not his decision to love her but his desire to forestall K that eventually forced him into a proposal. We can say that in this sense the basic motivation for the proposal derived from homosexual feelings.

His inability to forget his dead friend made Sensei apprehensive. "I tried to bury my anxiety in books. I began to read relentlessly." But books could not save him, and he gave them up. He spent his days in idleness and did not look for work. Fortunately his inheritance made this possible. There was also a period when he tried to drown himself in drink. That, too, proved unsuccessful. The thought of K's suicide dominated his life. At first he decided that K had died of a broken heart. "But later when I was able to think more objectively, I found myself considering whether that wasn't too simplistic a diagnosis. To say it was a clash between his ideals and reality also seemed inadequate." Finally, he considered the possibility that K had experienced a loneliness as terrible as his own and, "wishing to escape it quickly, had killed himself." His blood ran cold at the premonition that he might be following in K's footsteps.

His mother-in-law's illness, and the opportunity to care for her, offered a respite. When she died and he and his wife were left to themselves, his anxiety returned with renewed vigor. At times he was assailed by a "nameless fear." At first it came from outside without warning. Later, "I began to wonder if it had been hidden in a corner of my heart ever since I was born . . . I doubted my sanity." The sinfulness of man weighed on his conscience and he "welcomed a flogging at the hands of a stranger" or, better yet, at his own hands. No, he told himself, you must die. Finally he decided "to live as though dead," but no sooner had he reached this decision than he was overwhelmed by "a dreadful force." A voice castigated him—"You have no right to do anything"—and his frequent attempts to defy it came to no avail. "Why do you stop me?" he would ask, and with "a cruel laugh" it would reply, "You know very well why."

The mental state described here suggests that Sensei suffered from auditory hallucinations. When it became apparent that he could not escape them, he considered suicide. Still he hesitated out of consideration for his wife. Years went by, and finally the Meiji Emperor died. Sensei had been born and raised during his reign. An era had ended, and Sensei was greatly moved a month later when General Nogi committed ritual suicide and followed his beloved leader into death. Undoubtedly the story of how the general's banner had once fallen to the enemy, and how he had lived the last thirty-five years of his life thinking only of death, struck a familiar chord. He decided to die like Nogi. For the first time he found the inner freedom to execute his promise to his young friend. He completed a detailed testament and took his life.

I would like for a moment to consider the boy's reaction. On the basis of a reading of Sensei's letter one would have to judge him psychotic. This is not, strangely enough, the boy's impression. He never stops to think that the story about the uncle may be a delusionary experience. He has complete faith in his mentor and everything he says. It may very well be a naive faith, but then countless numbers of readers, including critics and Sōseki specialists, have also failed to recognize the true nature of Sensei's illness. In any event, I seriously doubt if the boy ever really understands why Sensei commits suicide. At the close of his letter Sensei writes equivocally, "Perhaps you will not understand clearly why I am about to die, anymore than I can fully comprehend the reasons for General Nogi's death." It is my impression that this statement represents a highly ingenious piece of camouflage. Sensei may insist that he had done his best to explain the enigma that is himself, but why does he then choose to obfuscate the most important point of his letter? Obviously he does not wish to spell out the idea that, like Nogi, he is loyally following his beloved into death. Or perhaps he prefers not to think about it. He is unable finally to express in writing his feeling that K is beckoning to him from beyond the grave.

Perhaps, as a man about to die, he feels a need to impart significance to the last moments of his life. We might call this his final struggle against death. On one occasion during the month that elapses between the Emperor's funeral and Nogi's, he tells his wife that, with the end of the Meiji era, they will become anachronisms. She suggests in jest that he try ritual suicide. "I will, if you like," he replies half seriously, "but out of loyalty to the spirit of Meiji." More likely than not, Sensei prefers to have his suicide understood as a sacrifice to a passing age than as loyalty arising from homosexual feelings for K. The passage, moreover, at the beginning of his letter should also remind us of the scene in which Sensei discovers K's body. I am referring to the passage quoted earlier in which he states his hope that his life will be re-born in the heart of his young friend. It seems to represent his attempt to stun the boy as he himself had once been stunned by K's death. Only this time the shock is intended to lead to "new life." Sensei hopes that his death will give the boy freedom and independence. He does not want the boy tied to him as he had been tied to K; and he presents his past as a final admonition. Unfortunately, the darker aspects of his life make almost no impression. Instead the boy views their recitation as proof of Sensei's agonized love. This is the reason he later describes his mentor "as a man capable of love, or rather as one who was incapable of not loving—who, at the same time, was unable to embrace the love others offered him."

Indeed, there is an obvious defect in Sensei's manner of loving. Perhaps it is love that prompts him to leave a testament for his young friend and follow K into death. But how are we to understand his adamant refusal to unburden himself to his wife and his insistence that the letter be withheld from her? Is it love that dictates that her memories be kept as unsullied as possible? No doubt Sensei loved his wife with all the affection he could muster, but it is also undeniably apparent that he assumes instinctively a posture of self-defense in his relationship with her.

I do not wish to impugn his sincerity. As he himself says, "It is a sin to love. And it is also sacred." He knew from his own wretched life how the desire to love could drive a man to evil. In spite of everything, however, he refuses to rule out the possibility of its sanctity. He believes in love although he has abandoned all hope for himself. This is the reason he respects the earnestness of his young friend's friendship; and why, as its reward, he warns the boy against respecting him as a mentor. Insane he may be, but on this point Sensei imparts a very important message. Indeed it is a very sane one.

Chapter Nine

MICHIKUSA (Grass on the Wayside)

1915

Synopsis

This novel is a fictionalized version of Sōseki's life. The central character, Kenzō, can be thought to be equivalent to Sōseki, and the action of the novel corresponds to the years 1903–1905, or the time of Sōseki's return from London up to the time his writing career was about to begin.

At age two Kenzō was given for adoption to the Shimadas, a childless couple who served his father. He lived with them until age eight when Shimada's wife divorced her husband on grounds of adultery. He was returned to his real family, but it was not until twelve years later that he was legally reinstated because of a row that ensued between his father and Shimada over who should bear the cost of his upbringing. These were bitter years, and he escaped them only by excelling at his studies.

After graduating from the university, Kenzō left Tokyo, his birthplace, and taught school in the provinces. He married Osumi, the daughter of a former Cabinet member; and they had two girls. Then the government sent him abroad to study for two years. He has returned to Japan recently to assume, at age thirty-six, a professorship at Tokyo Imperial University.

Kenzō doubts the worth of having become a scholar. He is a prisoner of his studies, and his salary of 120 yen a month is insufficient to his family's needs. Finding and furnishing a new house has consumed his savings; and a third child is on the way. There are no relatives to whom he can turn. An elder brother, Chōtarō, slaves at a lowly position in a government office, and a half-sister, Onatsu, unable to depend upon her husband, Hida, comes to Kenzō for financial help. Osumi's father has fallen on hard times.

Osumi does her best at economizing and even pawns part of

her trousseau to make ends meet. But what she cannot cope with is Kenzō's continual ill humor. Although a devoted couple, they have no ability to communicate. He closets himself in his study while she seeks the companionship of the children.

One morning on the way to the university, an old man crosses Kenzō's path. Kenzō recognizes him immediately as his former foster father. Although he has not seen him for sixteen years, the memory of this arrogant, grasping man is still vivid in his mind, and he quickly looks away. A week later when their paths cross again, Kenzō has a premonition that he will hear from Shimada. He says nothing to Osumi, whom he tells little of his childhood. The following Sunday he visits his sister Onatsu to learn what she knows of Shimada's present circumstances. She warns him to ignore the man. "He's the type who'll take the cooking pot out of your kitchen."

Sure enough, a man named Yoshida Torakichi calls at the house on Shimada's behalf. Kenzō is sick in bed, but Yoshida returns a second time. Will Kenzō see his aging father? Osumi, eavesdropping in an adjoining room, is distressed when Kenzō finally agrees. She predicts that sooner or later they will be asked for money.

Shimada calls with Yoshida a week later. In the meantime he has contacted Onatsu's husband, Hida, and proposed that he readopt Kenzō. Everyone is astounded at the ludicrousness of this offer, and Hida takes responsibility for saying no. Shimada calls a second and a third time. He and Kenzō have little to discuss, and Kenzō becomes annoyed at the way his time is wasted. Yet each visit releases a freshet of memories, and he is transported back to his childhood. He remembers clothes, toys, and fishing trips—as well as the violent arguments between the Shimadas in the last year of their marriage.

On the fourth visit, Shimada asks for money. Kenzō hands him the few bills in his wallet, which he leaves in the sitting room. Osumi, trying for once to spare his feelings, replaces the money and returns the wallet to his study. Kenzō says nothing, even though he wonders if she is making a fool of him, for she is now heavy with child. She is convinced that this time she will die in childbirth and, one night, Kenzō finds her in a hysterical trance clutching a razor.

Shimada continues to call, asking for larger sums. Kenzō learns that Shimada and his second wife, Ofuji, are living off a stipend provided by the husband of Ofuji's daughter, Onui. Onui is said to be dying of a disease of the spine.

Who should appear next at Kenzō's door but Shimada's ex-wife, Otsune. Kenzō gives her carfare; no sooner has she left than Osumi's father arrives to ask Kenzō to back a loan.

A third daughter is born one cold night before the midwife can arrive. Kenzō helplessly wraps the baby in cotton; Osumi survives without mishap.

The end of the year is at hand when Shimada calls again. Onui has died, and her husband has cut off his stipend. He needs a large sum to settle his year-end debts. Kenzō refuses, but several days later a representative appears from Shimada. At the time of the annulment of the adoption, Kenzō wrote an innocent note stating that, official matters notwithstanding, he wished to continue to see Shimada. If Kenzō will purchase this embarrassing letter, Shimada will trouble him no more. Kenzō realizes that he is being blackmailed but, despite poor health, he grinds out a manuscript to raise the money.

In the final scene, Hida and Chōtarō arrive at his house with a receipt for the money and the original letter. They join Osumi in rejoicing that the matter is concluded forever, but Kenzō remarks bitterly that nothing in life is ever settled.

Kenzō's Homecoming

We often hear of the difficulties of living abroad. It is no easy task to live in a totally different culture and to operate in an unfamiliar language; at times the stresses are great enough to upset a person's emotional balance. We seldom hear about the difficulties of readjusting to home, however. Fortunate is the man who returns to find home as he thought it would be, if not better. For the vast majority of people living abroad, home becomes either glorified out of proportion or blocked out of mind entirely. In the former case, homecoming means a shattering of illusions; in the latter, a confrontation with the past. In either event, it constitutes an identity crisis. The person will be forced to ask anew, "Who am I?" Unable to answer the question, he may suffer an emotional illness far more serious than any passing sense of dislocation experienced abroad.

This is the crisis that Kenzō faces when he returns from abroad and sets up house in Komagome, Tokyo. "Latent to the novelty he experienced at living in his native city again was a certain loneliness. The aura of the distant land he had newly forsaken still clung to him, and he hated it. He wanted to be rid of it, the sooner, the better. What he failed to recognize was the secret sense of pride and self-satisfaction he drew from the fact of being different." Rather than being a sentimental occasion for Kenzō, homecoming is a source of strangeness and loneliness; and, just as an emigrant is apt to deny his original heritage in order to adopt a new one, he is anxious to rid himself of the airs adopted overseas. In other words, Japan is Kenzō's home in name only, and it does not differ appreciably from any other country in the world. Kenzō does not know what it is to have a real home. This void probably always existed in his life, and it persisted despite his frantic efforts to lose himself in his studies abroad. Coming home reminds him only too well of his utter loneliness.

Kenzō is in his middle thirties. He immediately goes to work at the university, putting his time and energy into his teaching. His lectures are far more important than the wife and children he has not seen for some time, and everyday he retires to his tiny study as soon as he returns home and changes his clothes. "He felt continually weighed down by the mountains of work that awaited him in his little six-mat study . . . The truth of the matter was that he was motivated not so much by a desire to work but a compulsion . . . His mind no longer knew what it was to rest . . . This naturally required that he cut himself off from the company of other men. The more preoccupied he became with books, the more isolated he became as a person. There were times when he faintly perceived the extent of his loneliness, but he placed his confidence in the extraordinary passion that burned deep within him and fired him on. Thus, he knowingly turned his steps in the direction of desolation. The bleaker the terrain became, the more convinced he was of having found his true path in life." The point to be drawn here is that Kenzō's studies continue to be his sole reason for living. Perhaps they had been a viable means of dealing with loneliness abroad, but at home they have become a last, desperate resort.

Kenzō's relatives regard him as eccentric. He attributes this to the differences in their educational backgrounds; he was, in fact, the only member of the family to have gone to higher school. Of his family, only an elder half-sister and an older brother are still living. She is an asthmatic married to a ne'er-do-well office clerk; he is a petty civil servant. Both are in perpetual financial straits, and Kenzō is often obliged to supplement his sister's income on the sly. Consequently he sees them only a few times after his return and does not maintain close contact with them. He knows better, but then his work is more important. His wife, Osumi, finds his attitude pompous; and, in turn, Kenzō is "bitterly frustrated with a wife who did not understand him. At times he scolded her; at other times he berated her into silent compliance." His anger merely serves to substantiate her claim, and he is left to fume. Although he pays no attention to his relatives, he cannot ignore his wife's criticisms.

An unexpected event happens at this point, and it greatly upsets Kenzō. An old man, a Mr. Shimada, passes him on the way to work. Shimada was once Kenzō's foster father. The two have not met, however, since they severed relations sixteen years earlier. Shimada recognizes Kenzō immediately and stares at him, but at this encounter, as well as a second one several mornings later, Kenzō refuses to acknowledge the old man's presence. When he returns home he experiences "a strange premonition . . . that this is not the last of the matter." He says nothing to Osumi, however. On the following Sunday he decides on the spur of the moment to visit his sister to learn what she knows. She cannot help him, but that week a Mr. Yoshida calls at the house claiming to represent Shimada. Kenzō is sick in bed and cannot receive him. Later, Osumi mentions the caller and asks if Kenzō has any intention of seeing him. He consents when she urges him against it. "She attributed his decision to his usual egotism. Kenzō decided that, unpleasant as it may be, etiquette required he meet the man."

Kenzō sees Yoshida and, while he refuses to provide financial assistance, he agrees reluctantly to meet his foster father. This is the beginning of Shimada's many visits. Osumi is distressed by this turn of events, and she visits Kenzō's brother to learn why the two men had originally parted ways. It seems that Kenzō had been adopted by Shimada only to be later returned to his real parents. Kenzō is, according to her reasoning, under no filial obligation to Shimada. Shortly thereafter a postcard arrives asking Kenzō to join his brother and brother-in-law for a talk. Shimada has proposed readoption. No one favors this, of course, but Kenzō is indebted to his brother-in-law for acting as a go-between. Thus, a chance encounter with the foster father leads to a renewal of relations with the whole family—as well as a slight change in Kenzō's emotional equilibrium. It is no longer possible for him to forget the mundane world about him. A phase of his life that has been locked away in the past "suddenly had the uncanny ability to make itself very present."

Although Kenzō prides himself on having risen above his relatives, he wonders if the price has not been too great. He is reminded of the story of a woman released from prison after serving twenty years for murder, and he likens the time and effort that have gone into his academic career to her imprisonment. "His life so far had been built upon the years he had spent at his own version of hard labor; and his present toils were, he desperately wished to believe, very necessary to his future. This was the course he had charted for himself and, as far as he could see, it was the best possible. Yet at times he suspected that his life was nothing but a long process of aging." The terrible thought slips out: "You study all your life and then you die—what is the point?"

Kenzō loses the conviction that he is different from the rest of his family. When he sees his hapless brother, he wonders if he himself is not about "to be dragged in a direction contrary to the one he should be moving in ... He was pessimistic enough to believe that he could fall into the same straits as his brother." As for his foster father, "he was simultaneously a specter from the past, a living human being, and an ominous shadow of the future. It was not curiosity but anxiety that prompted Kenzō to ask how much of Shimada's ill luck had accrued to himself as well." A voice inside tells him that there is little difference between himself and his greedy foster father; and that, stripped of his education, he would be as ill-bred as his sister. In other words, Kenzō's fateful encounter raises not only bitter memories of the past but also questions of identify and raison d'être. He finds himself in the midst of an identity crisis.

He does not know how to handle it. Meanwhile Shimada's visits grow more frequent, and Kenzō is left to wonder what the old man wants. Osumi, equally afraid, loses all patience with her husband's passivity. She always tended to hysteria, and the ever-increasing frequency of her outbursts throws Kenzō into further anxiety. He loses the ability to absorb himself in his work. He begins his lesson plans only to abandon them. "I quit. I don't care anymore." Then, as Osumi predicted, Shimada asks for money and Kenzō cannot refuse him. On top of this, Osumi must tell her husband that she is pregnant. "There were times when he thought about money. There were even times when he asked himself why he had not chosen wealth as the goal of his life." Not that he has any genius for making money, but "he was beginning to be tired of being neither rich nor famous." His powerlessness makes him angry. "He was so touchy that he sometimes imagined he'd go mad unless he discharged the almost galvanic anger inside him. Once he kicked a pot of flowers from the verandah. It belonged to the children. They had begged their mother to buy it. Seeing and hearing the red pot shatter as it hit the ground made Kenzō feel better. But the sight of the broken stems and wounded flowers made him regret for a moment the cruel stupidity of his own actions. These poor flowers, he thought, were lovely in the children's eyes; now they have been ruined by the children's own father. He felt sorry for what he'd done, but unable to talk about it to the children. As usual, he was able to make excuses; 'I'm not to blame. Someone else is to blame if I act like a madman.'" Kenzō has fallen into a state that borders on melancholia.

Kenzō's Upbringing

Kenzō was adopted by the Shimadas when he was two years old. They had no children and asked Kenzo's father, their benefactor, to give them a son. At age two Kenzō must have been old enough to begin to comprehend the world about him. The encounter with Shimada reawakens many of these memories, and he recalls first of all the house where he lived. He remembers clothes and toys and a fishing trip. He cannot recall his feelings, however. Later he is able to recall the appearance and surroundings of two other houses with considerable vividness, but, as in the previous case, there is "not even a shadow of a human figure." This paradox greatly troubles him.

In psychoanalytic parlance, this type of vivid–but fragmented and dispassionate–memory is known as a screen memory, suggesting that a traumatic event has been repressed or screened out. One can imagine that adoption came as a shock to two-year-old Kenzō. Recent psychoanalytic observations suggest that an extremely small child becomes acutely melancholic when separated from the object of its love, and it is unlikely that Kenzō was an exception to this rule. A child of two may not possess strong feelings of sadness or loneliness but it can experience a kind of depersonalization. This is why perhaps people and feelings are so conspicuously absent from Kenzō's memories.

Kenzō is able to recall with considerable veracity mental pictures of the Shimadas and his feelings about them from a slightly older age. Part of the house served as a ward office which his foster father headed. Kenzō had free run of the building, and the office staff doted on him and his pranks. Shimada tolerated such naughtiness, and no one dared to correct the boy. The Shimadas were also unusually generous, for in reality they were quite miserly–a fact

Kenzō was to learn much later. They dressed him in expensive kimonos and granted his every request.

Undoubtedly their generosity stemmed from their anxiety. They persisted in asking, "Who is your father? Who is your mother?" and were delighted when he pointed to *them*. "It was awful for Kenzō. Sometimes he was more enraged than hurt and stiffened up, refusing to give them their answer." In his childlike way Kenzō probably realized that the Shimadas were not his parents; he perceived their insincerity intuitively. The better he was treated, the more confined he felt, and he instinctively revolted. "Though he loved his toys and spent endless hours studying his colored prints, he had no smiles to offer those who bought them for him. He wanted to enjoy a gift as a gift and to forget that it had any connection with its donor." He began to throw frequent temper tantrums and to insist on having his own way. "It seemed that all the world existed solely to take his orders."

Meanwhile Shimada and his wife, Otsune, began to have serious marital problems. Kenzō awoke one night to find them arguing furiously and, in confusion, he burst into tears. The arguments continued night after night until finally Shimada left home. Kenzō was able to gather from Otsune's tirades that Shimada had been sexually unfaithful. Still "he liked Shimada more than Otsune, who was forever trying to get him on her side. She would say, 'Remember. I have no one but you now. You must never let me down.'" Her attempts to attach him made her appear all the more odious.

Kenzō was returned to his real parents by age eight. "In the eyes of his father he was a little nuisance." The man displayed no interest in his son and made it known that he begrudged every cent spent on him. "To his father and Shimada, Kenzō was not a person—only property. To the former, he was unwanted goods; to the latter, he was a kind of investment that might later prove profitable." Technically, however, Kenzō remained Shimada's child, and he visited his foster parents occasionally. On one visit Shimada ordered him to find work as an office errand boy. Nothing could be more humiliating, and thereafter Kenzō applied himself rigorously to his studies in an effort to carve out an independent career He was twenty-two before his real father consented to make compensation for his education and upbringing, thereby absolving all ties with the Shimadas.

We see that Kenzō's childhood was conspicuously lacking in love. Kenzō did not really know what it was to be loved by a mother or father. The novel does not, admittedly, touch on the subject of his real mother, but her absence suggests her influence was minimal. In any event, Kenzō was raised in a loveless environment and at one point in his life he decided to escape it on his own merits. His efforts seem to have been sufficiently rewarded. He has studied abroad, and he has established himself as a university lecturer. His success is a source of pride, but it is also the cause of his isolation. Can he continue to live this way? The recollection of his past "imprisoned life," the pattern of his feelings toward his wife, and the constellation of feelings aroused by his encounter with Shimada suggest that he cannot.

"Nothing's Ever Settled in Life"

I have already mentioned that Kenzō is deaf to all criticism except his wife's. This suggests that he secretly expects her to take his side. He expects her full support, although he exerts no effort to explain his feelings to her. This is an extremely infantile attitude and it manifests itself in nearly all aspects of their relationship. In one incident, when he is sick in bed for several days with a high fever, he awakens to find Osumi sitting by his pillow looking very worried. He realizes she has nursed him back to health and that he is fortunate to have a loving wife, but he turns his head away in silence. He probably thinks that an expression of gratitude would sound affected. It is only natural that, as his wife, she should care for him. He appreciates the readiness with which she ministers to him, but he resists any thought of being expected to thank her. Consequently he rejects any special attempt on her part to make him happy. In another incident, she shows him a length of cloth she plans to tailor into a kimono for him. Her face is radiant with happiness, but "in Kenzō's eyes her smile seemed to conceal a poor attempt to get on his good side. He questioned her sincerity. And he deliberately pretended to be unimpressed with her thoughtfulness."

Kenzō's attitude toward his wife clearly reflects the pattern of his childhood experiences. He expects the unconditional love that parents give their offspring, and if she fails to provide it he revolts much as he had done with the Shimadas. He is not fully aware of the nature of his own behavior, and Osumi is even less comprehending. She thinks him either cold or excessively demanding. Then, too, just as Kenzō wants Osumi to be the mother of his dreams, she wants Kenzō to be a man like her father. This conflict accounts for the high degree of tension that exists between this couple, and sometimes it manifests itself in the form of Osumi's paroxysmal trances. Kenzō is fully cognizant that his wife is behaving hysterically, but he is not one to sit idly and let her stare into space. In an almost prayerful manner he would grasp her by the shoulders and try to talk to her, or, when she slept too long, he would anxiously awaken her. His actions resemble those of a child upset because it cannot rouse its mother.

We should be able to detect the same pattern in the mixed set of emotions Kenzō entertains about Shimada. On a conscious level Kenzō hates his foster father and does not wish to see him. Still he cannot reject his request for a meeting even though his wife and brothers are convinced nothing good will come of it. His excuse is that he is indebted to Shimada for having raised him. There are, however, other passages which suggest a different motive.

Unlike his brothers, Kenzō is incapable of taking Shimada's offer for readoption lightly. He finds it genuinely puzzling. Why would Shimada want to see him even though he has been refused financial help? From Osumi's point of view, this is a truly silly question. The answer is obvious—relations had better be broken off now. Kenzō cannot agree. "There is a great deal in this world that one cannot dismiss simply because it is a nuisance." This passage seems to suggest that, whether Kenzō is fully conscious of the fact or not, he entertains secret expectations about Shimada.

He wants to enlist his aid in unraveling their past relationship and in establishing a new adult one. He wants, if possible, to establish warm and genuine communication with him.

Shimada is too insensitive to appreciate Kenzō's feelings. He becomes brazen and asks for money outright. We might join Osumi in chiding Kenzō for not having known better, but that would be to ignore the pain he carries in his heart. This is the pain that he later feels, for example, when his foster mother comes to visit. He is saddened by his inability to empathize with her for she has become a shadow of her former self. He would cry if he could. He gives her money without being asked and says to himself as she leaves. "Had she been more decent I might have wept for her. Or, failing that, made her a little happier. I would have lifted her out of poverty and been with her when she died."

Since no one can appreciate this feeling, Kenzō has no choice but to proceed without the understanding of his wife and brothers. "I told you so," says Osumi when Kenzō becomes discouraged with Shimada's capriciousness; and, when he announces that the relationship can be terminated at any time, she says reprovingly, "But then everything you have done will be wasted!" Kenzō is sorry that his wife worries only about money. "To an outsider like you, the Shimadas may mean nothing; but I see things differently." Taking his criticism poorly, she retorts, "Well, in your opinion, I'm a fool." Kenzō makes no further attempt to change his wife's mind. Later Shimada asks for a fairly large sum of money on the grounds that his second wife's daughter has died leaving him without any future means of support. For the first time Kenzō loses his patience and flatly refuses. Shimada leaves in a huff, but two or three days later he sends word that he will never trouble Kenzō again if he is compensated for severing their relationship permanently. Out of a sense of love and obligation, Kenzō promises to raise the money because Shimada is in real financial trouble. Thus he has to scrape together a large sum at a point when his finances are most strained. When Osumi complains, he says boldly, "I do not have to help him, but I will." While these words reflect Kenzō's deep concern for his foster father, Osumi interprets them as additional proof of her husband's stubbornness.

Despite his poor health Kenzō grinds out another manuscript and raises the necessary sum. He gives it to Shimada and receives in return a formal letter of separation. Kenzō's wife and brothers are jubilant now that the matter is settled. Kenzō reminds them, however, that only an outward settlement has been achieved and that, in fact, nothing has been fundamentally resolved. "'Nothing's ever settled in life. Events repeat themselves endlessly. But they turn up in various new circumstances and we are fooled.' His words were bitter, as though his gorge were in his throat. Osumi said nothing. She picked up the baby, 'Nice baby, nice baby,' she said kissing its rosy cheeks again and again. 'You and I don't understand Daddy, do we?'"

Kenzō's concluding statement is extremely meaningful. He seems to be saying more than "history repeats itself"; he is speaking of an inner truth that he himself has experienced. As we have already seen, his experience with his foster parents sets the pattern for his later life and even affects the nature of his marital relationship. Twenty years later he is attempting to obtain from Osumi and the Shimadas what he could not gain as a child. His interpersonal relations are characterized by transference and repetition compulsion. He feels compelled to act out his past in order to erase it. His remarks seem to suggest, however, that he has overcome this compulsion because the recognition that nothing is ever settled implies an understanding of the impossibility of wiping out the bitter experiences of the past. If Kenzō truly understands this, perhaps he will no longer need to act out. Earlier, a voice inside of him asks what he is meant to do with his life. He shouts that he does not know, but the voice mocks him and counters, "Oh yes you do. You know very well—but you can't achieve it, can you? You've allowed yourself to become sidetracked." Kenzō avoids the issue, claiming, "It's not my fault." But if we assume that he has become aware of the inanity of compulsive behavior then this fixation should begin to disappear. His past anxieties will become "idle distractions" (*michikusa*),[21] and he can turn with renewed courage to his journey through life to its final destination, death.

Chapter Ten

MEIAN (Light and Shade)

1916

Synopsis

The Tsudas, Yoshio and Onobu, are unhappy after six months of marriage. Yoshio, a twenty-nine-year-old salaried worker, plays the role of doting husband but he would have preferred to marry Kiyoko, a quiet beauty who rejected him for another man. At age twenty-three, Onobu seems an obedient wife, but Yoshio is not the man her intuition divined him to be. They manage to live together by indulging each other's whims. This has been possible because Yoshio's father, who lives in Kyoto, supplements his newlywed son's salary. Yoshio reneges on repayment, however, and this month there is no check in the mail. To make matters worse, he is faced with a second operation for his hemorrhoidal condition.

Although Onobu's aunt and uncle, the Okamotos, can afford to help, Onobu is reluctant to ask. She is indebted to them for having raised her and provided a handsome dowry. Yoshio also has an aunt and uncle—the Fujiis—who raised him, but his uncle derives little income from the small magazine he publishes. A comfortably married sister, Ohide, refuses to help Yoshio because she is angry with him for not repaying their father.

Yoshio decides to proceed with the operation the following Sunday, a day originally planned for an outing to the Kabuki theater with the Okamotos. Onobu declines the invitation, but no sooner is Yoshio back from the operating room than she leaves him to join her family.

The occasion is a marriage interview for her cousin, Tsugiko. It has been arranged by Mrs. Yoshikawa, wife of the president of the company where Yoshio works. She acted as go-between for Yoshio's marriage to Onobu and, unknown to Onobu, for Yoshio's abortive relationship with Kiyoko. Her presence reminds Onobu of the strange rapport this woman enjoys with her husband. Feeling uneasy and hostile, Onobu retires from the conversation at an after-the-theater dinner party.

She bursts into tears, moreover, when she is invited home and the next day her uncle presses for her opinion of the groom proposed for Tsugiko. He feels that his niece has special powers of intuition in such matters. The Okamotos are shocked at her tears and wonder aloud at how her personality has changed. She is too proud to explain and, in a conversation with Tsugiko, she perpetrates the fiction of her happy marriage. When she leaves, her uncle presents her with a check to compensate for having made her cry.

As she dresses the next morning, a Mr. Kobayashi appears to collect an old overcoat promised him by Yoshio. Kobayashi works as a proofreader for Yoshio's uncle and is about to leave for Korea to assume a new job and, according to him, to escape police surveillance for his socialist sympathies. Yoshio asked him specifically not to come while he was away, but Kobayashi derives great pleasure from needling people. Finding Onobu alone, he uses the occasion to insinuate that Yoshio keeps her in ignorance of his past romances.

The same morning Ohide arrives at the hospital with money. She attempts to extract an apology for Yoshio's lack of consideration of their father and, just as she is making a vague remark about his unresolved feelings for Kiyoko, Onobu enters the room. Onobu's suspicions are now thoroughly aroused, but she produces the check from her uncle so that Yoshio will be in a position to reject his sister's aid. Chiding the Tsudas for their selfishness, Ohide deposits her packet of money and leaves. Husband and wife in victory share a rare moment of togetherness.

Kobayashi appears at the hospital the following day. Yoshio is unable to learn what Kobayashi has said to Onobu, and he is even more distressed to hear that Ohide has been to complain not only to Uncle Fujii but also to Mrs. Yoshikawa, who is on her way to see him. He promises Kobayashi money if he will leave before she arrives, and he dashes off a note telling Onobu to stay home.

Onobu has, however, already gone to Ohide's house, ostensibly to effect a rapprochement but also to inveigle her sister-in-law into telling Yoshio's secret. The conversation proves fruitless.

Meanwhile, Mrs. Yoshikawa is stating to Yoshio her agreement with Ohide: Yoshio wastes money spoiling Onobu because he is still in love with Kiyoko. She proposes he see Kiyoko one last time. He is to travel at her expense to a hot spring spa where Kiyoko is recuperating from a recent miscarriage.

Yoshio is discharged several days later and, as promised, arranges a farewell dinner for Kobayashi at an expensive restaurant and presents him with thirty yen. Kobayashi is unimpressed and lectures Yoshio on his class consciousness. He invites a young, indigent artist, whom he had wait outside, to join them. Next, he has Yoshio read a letter written by an anonymous student in desperate need of money. Spreading the three ten-yen notes from Yoshio on the table, he offers them to the artist.

Yoshio arrives at the spa with great misgivings. Before he can decide to abandon Mrs. Yoshikawa's plan, he accidentally encounters Kiyoko, who makes a frightened and hasty retreat to her room. The next morning he has a basket of fruit delivered to her as a gift from Mrs. Yoshikawa. She is puzzled by this unexpected thoughtfulness and receives him to inquire about it. Their conversation proceeds desultorily and, surprised at her ability to have regained her composure, he fails to broach the question of why she rejected him. The novel ends, unfinished, with Yoshio ruminating over the meaning of Kiyoko's enigmatic smile.

Yoshio and Onobu

The characters in this novel do not evince the types of abnormality found in the novels after *Sanshirō*. They would be considered normal by today's social standards, and Sōseki proves himself equally the master of characterization of the average adult.

Tsuda Yoshio is still feeling restless although he has been married for over a half a year. His intuition-oriented wife, Onobu, strikes him as too formidable a mate. He had previously been involved with a different type of woman, Kiyoko, and Mrs. Yoshikawa had acted as an informal matchmaker. It seemed inevitable that the two would wed until one day Kiyoko suddenly abandoned him for another man. This was a bewildering experience, and Mrs. Yoshikawa, feeling responsible, offered to act as official go-between when Onobu's name was proposed as a possible bride. Her husband was both the president of the firm where Yoshio worked and a friend of Yoshio's father. Onobu was the niece of his friend Mr. Okamoto, who raised the girl like his own child. Yoshio had no objection to the marriage; in fact, he welcomed it as a means of insuring his future success.

In other words, we can think of Mrs. Yoshikawa as his patroness. "In one sense Yoshio enjoyed being treated like a child by this woman because of the special air of intimacy that such treatment produced between the two. It was a kind of intimacy that, scrutinized closely, one would find unique to heterosexual relations. It was like a pleasant sensation a man experiences when a geisha chummily pokes him in the side." Yoshio prides himself, nonetheless, on his strong ego, and he is confident that he cannot be made to behave like a child. He keeps this aspect of his feelings hidden; and "thus, while on the surface he casually tolerated her unreserved teasings, underneath he leaned on the thick, heavy wall that he had built against her."

In a similar manner, Yoshio's aunt often takes her nephew to task as though he were a child in need of a reprimand. When, for example, he attempts to attenuate her criticism of him as "too extravagant" by gratuitously agreeing with her, she goes on, pointing to a fundamentally restless streak in his personality. "It's not what you eat or what you wear. It's the way you're constituted. You are like a man who is forever on the lookout for a free lunch." When he suggests facetiously that that puts him in a class with beggars, she chides him for a lack of "natural seriousness." "It is a fine thing when a young man learns to accept what he is given." And, in a remark implicitly directed at him, adds that "there are people so particular about their choice of mate that even after the wedding they are too choosy to settle down."

Onobu, too, gradually becomes restive at the thought that

her marriage is not what she anticipated. She asks herself if "a husband is a sponge existing solely to absorb a wife's affections," and she blames herself for not being able to handle him better. "There are younger women who can manipulate men infinitely more temperamental than Yoshio. I must be rather brainless if at twenty-three I can't manage him at will." On one occasion she detects from Mrs. Yoshikawa's attitude that some fact relating to Yoshio's past has been concealed from her and, when Kobayashi's insinuations and Ohide's remarks tally with these suspicions, she becomes convinced that the secret concerns a liaison with another woman. She cannot bring herself to reveal her fears to her aunt and uncle, however. This is partly a matter of face and partly her desire not to destroy their image of her as a happily married woman. Further examination will also reveal how her complex feelings for her uncle influence this decision.

Onobu secretly prefers her uncle, who is a relative by marriage, to her aunt. "She believed that in exchange for her favor he paid her special attention . . . She could please him with almost no effort and feel gratified at the same time . . . Since her training in handling the opposite sex derived solely from her experiences with him, she believed that any marriage would succeed if she applied the same techniques." Her life with Yoshio soon exposes the fragility of this expectation, and she is forced to recognize that she can only relate to men like her uncle. Much as she would like, she cannot bring herself to confide in her uncle because she perceives that he has never really approved of her choice of a mate. His question on one occasion concerning whether she genuinely likes "Yoshio's type" leaves her at a momentary loss for words because it seems to imply an automatic dislike for him; and there are even times when she gloats at the thought that her uncle may be jealous of Yoshio. Because of these complex feelings Onobu fears that to unburden herself to her uncle is to somehow give way to a dangerous impulse.

On the day that Yoshio is admitted to the hospital, her uncle insists that she join the family at the Kabuki theater. When she is told the next day that the occasion was a marriage interview for her cousin, not only is she unable to give her impressions of the proposed groom but she also begins to feel resentful that her plainer looks were enlisted to show off her pretty cousin to good advantage. Her uncle is unsparing of her feelings and tries to wrest an opinion from her. "If any woman can discern a man's character in a single glance, it would be you," he says. Onobu imagines that her uncle is poking fun at her deliberately and, unable to tolerate the situation any longer, she asks, "Why do you have to tease me?" and bursts into tears. Mrs. Okamoto is visibly distressed by this display of emotion and rebukes her niece—"What is wrong with this child? She's not the type to fret. When she lived with us, she never cried no matter how much she was teased"—but Mr. Okamoto comes to Onobu's defense: "I was wrong to tease . . . That's it, isn't it, Onobu? I'm sure of it. Hush now, and I will give you something nice." Later he writes her a check.

So far we have examined the emotional status of the Tsudas' married life and the constellation of relationships surrounding it. Interestingly enough, we find that Yoshio and Onobu share a number of points in common. Both were raised by an aunt and uncle, and both are disproportionately attracted to the aunt or uncle who is their sexual opposite. The novel is not as explicit on this point with regard to Yoshio, but his nonchalance toward his aunt's unsparing criticism suggests that he expects she will continue to indulge his shortcomings and that he can entertain dependency wishes (*amaeru*) toward her. Mrs. Yoshikawa represents, moreover, a second older woman with whom he enjoys a special degree of intimacy. She has considerably more sexual appeal than his aunt, although the two women are approximately the same age. It appears that in tandem these women perform a role equivalent to the one Mr. Okamoto plays for Onobu. That Sōseki himself is aware of this point is apparent from the anecdote he has Okamoto tell at the dinner table on the evening of Onobu's visit. He has heard it from Yoshio's uncle, Mr. Fujii. "According to what the old boy says, it works this way. He says it is natural that the male offspring prefer the mother; and, conversely, the female, the father. You know something? He's right." This is, of course, what psychoanalysis calls the Oedipus complex. We can assume that because Yoshio and Onobu did not enjoy a close relationship with their parents they came to entertain strong oedipal feelings toward their aunt and uncle.

They are no exception to the fact, moreover, that the Oedipus complex also entails revolt against the parent of the same sex. Although Yoshio does not direct the kind of hostility he reserves for his father toward his uncle, he probably does not respect Fujii who, in his words, "has no experience in standing before the real world and dealing with cold facts." Onobu is likewise highly critical of her aunt. "There were even times when she wondered how anyone subject to so few ups and downs in life could be so rigid in her views." We are also told that she vows never to become inflexible in her old age. We can conclude that neither Yoshio or Onobu possessed in their surroundings a member of the same sex whom they cared to emulate. If, as psychoanalysis holds, normal adult development is contingent upon the child's eventual identification with the parent of the same sex, we can understand why the Tsudas, each lacking a fully socialized adult with whom to identify, have yet to discern what their respective roles in society should be. It is also possible to see this as the reason for their continued restlessness.

There are characters in the novel who do understand their place in society. Both the Fujiis and the Okamotos probably qualify; and Yoshio's younger sister, Ohide, is depicted as a mature representative from the Tsudas' generation. She is always her father's ally in the contest that ensues when Yoshio breaks his promise to repay his monthly advances. She is kindhearted enough, nonetheless, to offer financial support when her brother is hospitalized even if she cannot resist the temptation to remonstrate with him for his misdeeds. Her well-meant intentions come to naught, however, when Onobu produces the check from her uncle, so as a last resort she castigates her brother and sister-in-law for the way in which they think only of themselves. "Yoshio, you love only yourself. And Onobu here loves only to be loved by you. In your eyes, there is room for no one else. Not for your sister, or even your mother or father." Ohide hopes that by giving Yoshio money she will induce him to act in a manner befitting an elder son. Unlike her modern brother, who sees no contradiction in living for the gratification of his own egoism, Ohide lives in accordance with traditional Japanese morality. Perhaps the best proof of this fact lies in her almost unquestioning acceptance of her husband's philandering and her contentment with her place in life as his wife.

It is worth noting Onobu's reaction to Ohide's little sermon: "Could, by chance, Ohide be a Christian?" She later asks her sister-in-law in person, and one feels that her remark is indicative of the social role that Christianity has played in this country since the advent of the Meiji era. It is a fact that Christianity in the last hundred years has, unlike the days of St. Francis Xavier, served to preserve and reinforce moral and philosophical traditions rather than to revolutionize them. It has had the effect, consequently, of driving people who are discontented with established society and morality to those Western philosophies that are either unrelated or antithetical to it. This tendency is more pronounced than ever before. As a matter of fact, modern Japan is inundated with men who, like Yoshio, believe that morality consists of thinking of themselves first and foremost, and women who, like Onobu, are determined to show the world that they have found happiness in a perfect marriage. And while people who, like Kobayashi, have radical ideas no longer need to feign isolation and have taken to the streets in droves, the Ohides, who are content with their lot, are a vanishing species indeed.

"I Live To Be Disliked"

Kobayashi adds a new twist to the aunt's analysis of Yoshio's inability to be satisfied and settle down. Whereas the aunt blames Yoshio's restlessness on his penchant for extravagance, Kobayashi says, "Your kind of restlessness is itself an extravagance." He is about to depart for Korea with tenuous hopes of finding a better life, and he makes this remark when Yoshio responds sympathetically to his complaint that society will not let him settle down.

"It's not you alone, you know," says Yoshio. "I can't settle down in the slightest either." Kobayashi considers it a bit rich that a person in comfortable circumstances should complain this way, and he points out that it is economic freedom that makes Yoshio's restiveness possible. Once again Yoshio repeats his assertation: "Restlessness is a sign of these modern times. You are not the only one who is suffering." This means virtually nothing to Kobayashi who judges all matters in terms of his own circumstances.

Actually, Kobayashi is quite desperate at this point and is apparently feeling hostile toward his more affluent friend. Although he promises to collect the overcoat after Yoshio is discharged from the hospital, he calls at the Tsuda residence in Yoshio's absence and, in addition to boldly demanding the coat from Onobu, he unleashes a torrent of innuendo. He explains to Onobu why he enjoys annoying people. "Mrs. Tsuda, I live to be disliked. I deliberately say and do what people dislike. It is the only way I can tolerate the pain of living. I can't get people to pay attention to me. I'm a good-for-nothing. It is my aspiration, therefore, to make myself as obnoxious as possible because I know I shall never enjoy complete revenge for the scorn shown me." It appears that by provoking the scorn of people who belong to society Kobayashi is able to reconfirm the fact of his existence, however alienated it may be. At the same time, he derives a perverse sense of pleasure from exposing the weakness that underlies such scorn. "I would rather be alive," he says trying on the old overcoat, "no matter how much people laugh at the clothes I wear." "Really?" asks Onobu, "I think I'd prefer being dead to going through life that way." "In that case, Mrs. Tsuda," he advises her icily, "you should be particularly careful not to get yourself laughed at."

On the following day he invites himself to her husband's sickroom. He touches on the subject of Yoshio's argument with Ohide and, pouring on the sarcasm, he advises Yoshio that, whereas belligerence works to his advantage, "your case is different. Quarreling will never benefit you. And there are few men in this world who appreciate profit and loss as well as you." Yoshio anticipates that he will be importuned for money, and Kobayashi does force him into a promise for assistance. Later when Yoshio arranges to make payment, Kobayashi is presented with a second opportunity to make sport of him. He attacks along previous lines. "As I see it, you are always nervous and apprehensive. You are forever in avoidance of the unpleasant and in madcap pursuit of the things you like. Now ask yourself why. It's really quite simple—you possess more freedom than you can manage. You can afford to act extravagantly. You've never been reduced to my extremity where one accepts everything and lets everybody do as he damn well pleases." He points out that a great inner struggle is taking place inside of Yoshio as a result of his disappointment in marriage; and he has him read a letter written by a poor, anonymous student. When asked why, he explains as follows. "At least you felt a little sympathy for him, didn't you? That's enough, as far as I am concerned. It means that you both wish to give and to deny him the money. And that makes you uneasy because your conscience hurts. In that case, my purpose is fulfilled."

It would be safe to say that Kobayashi's real aim is to strip away the mask of nonchalance that makes Yoshio blind to his own vulnerability. As the two part, he realizes that Yoshio intends not only to ignore his counsel but also to prove it meaningless. "Very well then," he says, "Let's see who wins. The censure of experience may be far more swift and poignant than any lesson I can offer." This appears to be a hostile remark, but it is more on the order of the hostility that the strong express for the weak, than vice versa. Actually, it is more on the order of friendship than enmity. Kobayashi attempts to awaken his friend to the fact that, while they may live in different worlds, their ultimate destinies are the same and that Yoshio's position is all the more precarious because he believes it secure. Yoshio rejects this warning, however. If anyone is hostile, it is he.

The Unfinished Ending

When Yoshio is discharged from the hospital, Mrs. Yoshikawa arranges a change of scenery. She creates the makings of a little drama in which he and Kiyoko will meet at the same spa. Her theory is as follows—Yoshio's attitude toward Onobu appears strained. He makes, admittedly, an effort to be solicitous in order to maintain appearances, but in his heart he does not love her. This is because of his longing for Kiyoko. Therefore it would be best for him to utilize this opportunity to learn why he was rejected—Yoshio does admit to unresolved feelings even if he no longer feels love for Kiyoko. He agrees to Mrs. Yoshikawa's proposal, which we may interpret as her attempt to use him as a means to satisfy her own curiosity. Her failure to match the two had been no small blow to her ego.

Yoshio feels as though he is standing in a dream when he alights at the train station near the spa. "I am about to have the next episode of this dream that has haunted me since I left Tokyo or ever since Mrs. Yoshikawa suggested I come here. Come to think of it, it has haunted me since I married Onobu—and even that is not going back far enough. I have been cursed with it since the moment when Kiyoko turned her back on me. Here I am madly pursuing it again. I wonder if, after I have carried it this far into the present, it will suddenly disappear when I reach the inn. That was Mrs. Yoshikawa's idea. It must be mine too since I agreed and I am here to carry out her plan. But is she right? Will it suddenly vanish? Am I standing here in this dreamlike hamlet with no more faith in this plan than I now possess?—Yesterday was a dream, today is a dream, tomorrow will be a dream. And it will follow me back to Tokyo. I guess that is a kind of ending too. In fact, that is probably what will happen. Why did I leave the rain-laden skies of the city to come this far? Is it because I am, after all, a fool? I might as well give up now."

Yoshio is conscious of a similar confusion when he arrives at the inn. A voice says to him, "Do as you wish for the moment. Be a guest here for a cure, if you like. You are a free agent no matter what happens, and freedom means happiness. On the other hand, it resolves nothing; and that is why it is ultimately unsatisfactory. Will you therefore throw it away and, having lost it, grasp something in its stead? Are you sure of that? Your future is undetermined and it may pose far more riddles than the past. How will you fare in the struggle to unravel your past? Will you be proven wise or foolish in this attempt to cast aside your freedom and place the favorable resolution of your problems in the hands of the future?"

It is at this point that Yoshio chances upon Kiyoko. He is searching for the way back to his room from the bath as she steps into the corridor. He witnesses her body grow tense and her facial muscles tighten; her color turns noticeably white. Just when he feels compelled to speak, however, she turns on her heel and disappears into her room. After a sleepless night he sends the maid with a calling card, indicating his intention to visit. He notes on the card that an accompanying basket of fruit is a get-well gift from Mrs. Yoshikawa. In a few minutes he is ushered into Kiyoko's room, and his first impression is of a new formality—an irreconcilable distance has arisen between them. Kiyoko does not grow tense and pale as she had the night before. She thanks him for the fruit, and he asks after her husband. Presently their conversation turns to the subject of the previous evening. Yoshio denies having deliberately waited outside of her door and then, in what is apparently a strategem to draw out her feelings, asks how she could suspect him of such behavior. Kiyoko will not take the bait. She says simply, "It's only that I know you are the type of person who would."

He presses her with another leading question. How could she appear so calm this morning? Kiyoko makes no effort to reply and changes the subject to the matter of her surprise at not only having received a gift from Mrs. Yoshikawa but also having received it from, of all people, Yoshio. Sending the basket compliments of Mrs. Yoshikawa had been Yoshio's means of insuring that his request for an interview would not be questioned or denied. This deliberate foresight leads to an unwanted misunderstanding and he is forced to fabricate a second story: that he has come of his own volition and not as an errand boy. It happened that Mrs. Yoshikawa knew that Kiyoko was staying at the same inn and asked him to deliver the gift. "That must be it," replies Kiyoko, "Otherwise it doesn't make sense." And she cuts him short when Yoshio, feeling he has not explained himself fully, begins to offer other excuses, such as the possibility of coincidence. "I said it makes sense. Anything does if you ask and find out why." Yoshio, too, is tempted to ask why—why, that is, she had left him for another man.

Since the novel breaks off at this point we will never know if he does ask his question and, if so, what Kiyoko says. We might engage in a little conjecture. First, we might consider why Kiyoko is able to regain her composure so quickly. When Yoshio calls at her room the novel depicts her as appearing in the doorway dangling the basket of fruit at arm's length. This pose strikes him as typical of her brand of humor, but it is clear from her immediate expression of puzzlement at having been the recipient of such a gift that the pose is not meant in jest. She had been upset the previous night because she believed that Yoshio had followed her to the inn of his own accord, but on the following morning she has her wits collected well enough not to be taken aback by what is admittedly a most surprising gift. This is an extremely significant fact, for when Kiyoko realizes that Yoshio is hiding behind Mrs. Yoshikawa's name she no longer needs to fear him. And isn't this perhaps also the reason she had originally forsaken him? No doubt she had realized that Yoshio operated under Mrs. Yoshikawa's influence. She had been distressed by his dependence (*amaeru*) on this woman and by Mrs. Yoshikawa's peculiar interest in him. She probably also realized that Mrs. Yoshikawa would always have a say in the management of their relationship. We can imagine that this is the reason Kiyoko felt Yoshio to be unreliable.

There is no limit to the number of endings we might compose for this novel. Will Yoshio's encounter with Kiyoko awaken him from his dream? Will he, recalling Kobayashi's advice, realize that his old antagonist is really his friend? Will this realization enable him to settle down at last? Will he be able to reject a semblance of freedom for a life devoted to loving Onobu? And what about Onobu? Will she achieve new insight into life as a result of Mrs. Yoshikawa's plan to school her in the art of "the wifely wife?" We have no means of knowing. Perhaps by developing his characterization of Kiyoko, Sōseki planned to present his conception of the ideal woman or ideal human being. I cannot believe, however, that he has provided us with sufficient material to resolve the problem that he poses in the character of Yoshio. The statements of his aunt, Ohide, and Kobayashi make on the basis of their particular perspectives represent diagnoses of the problem, not solutions. Sōseki himself must have recognized that Yoshio's problem was not only an exceedingly serious one but also one that represented in compressed form the dilemma of modern man. We know from his correspondence that he was still searching for the meaning of life and that he was renewing his efforts to find the Way.[22] It appears, however, that when he realized he would be unable to find an answer to Yoshio's problem, and that the novel would have to be concluded without the solution he so earnestly desired, he collapsed from massive internal hemorrhaging and never rose from his sickbed to write again.

The very fact that death suddenly came to Sōseki as he was bringing his novel to a close strikes me as the ultimate answer to the anxiety of modern man as characterized by Tsuda Yoshio. Modern man has an aversion to thinking about death; he has made its discussion taboo. He knows that death is inevitable, but he lives as though it were not his problem. In this respect all of the characters—no matter whether polar opposites like Kobayashi or Yoshio—are the same. As Yoshio is leaving the doctor's office at the beginning of the novel he says to himself, "Who knows how and when this flesh might be subject to sudden change, let alone what changes it may be undergoing at this moment unknown to me. It is a frightening thought." This is his roundabout way of considering the topic, and it is apparently a reflection of Sōseki's unconscious premonition of his own rapidly approaching death. Instead of having Yoshio discuss the subject in explicit terms, Sōseki leads his thoughts in a different direction. "The same holds true for the realm of the mind. One never knows how and when it will change. And what is more, I have seen it change before my very own eyes." He appears to be referring to the dream that haunts his thoughts after his rejection at Kiyoko's hands. Wouldn't it be far truer to say, however, that he has been in pursuit of a dream—the dream of perfect happiness—long before he ever met her? Although his estrangement from her partially destroys this dream, he is still haunted by a sense of ethereal unreality, and even a second encounter with Kiyoko may not drive it from his mind. One can imagine that Yoshio's dream—the dream of modern man—has a never-ending quality that is dispelled only by confrontation with the grim reality of death.

NOTES

Introduction

1. Doi Takeo, *Sōseki no Shinteki Sekai* (Tokyo, Shibundō, 1969), p. 12.
2. Doi, *Sōseki no Shinteki Sekai,* pp. 240-241.
3. *Amae,* 甘え ; *amaeru,* 甘える .
4. "The concept of *amae* is indispensable to an understanding of the psychology of the characters that appear in Sōseki's works. Let me say that one of the reasons I undertook the analyses of these works is to prove this point. Admittedly, Sōseki himself does not use the word *amae* in dissecting the psychology of his characters, but it is quite clear that the concept of *amae* is a key to unlocking their psychology. Those who are already familiar with my writings will have undoubtedly picked up on this point, but let me summarize it briefly. First, Botchan appears to be a person devoid of any connection with *amae,* but we find that *amae* is latent to his relationship with Kiyo even though he is unaware of this fact. Second, Sanshirō retains a secret attachment for his mother at home at the same time that he deliberately turns his back on her and is drawn to the gay world of young love and sexuality. Here we see the truly naive face of his *amae.* The common denominator in Daisuke's relationships in *Sorekara* is also *amae,* and the novel hints at close interrelationship between *amae* and homosexual feelings. Sōsuke in *Mon* possesses a similar make-up, but in the case of Sunaga in *Higan sugi made* the frustrated *amae* goes a step further and assumes the form of a warped personality. The personalities of Ichirō in *Kōjin* and Sensei in *Kokoro* are characterized by distorted *amae* and homosexual feelings, both of which form the basis of their psychopathology. Kenzō in *Michikusa* can be thought of as Sōseki's self-portrait, and the novel is the agonizing account of the frustration of his wish to *amaeru* and of his unsuccessful attempt to conquer it. All of these characters suffer from frustration of their wish to *amaeru* and stand in contrast to the Tsudas, who are depicted in *Meian* as a couple who *amaeru* to each other. They are not, despite this capacity, happy human beings, however." Doi, *Sōseki no Shinteki Sekai,* Afterword, pp. 234-235.

5. Natsume Sōseki, *Sōseki Zenshū* (Tokyo, Iwanami Shoten, 1966), II, 417.
6. Doi, *Sōseki no Shinteki Sekai,* pp. 236–240ff.
7. Akagi Kōhei, *Natsume Sōseki* (Tokyo, Shinchōsha, 1917), p. 312.
8. Komiya Toyotaka, *Natsume Sōseki* (Tokyo, Iwanami Shoten, 1938).
9. *Sokutenkyoshi,* 則天去私.
10. Etō Jun, *Natsume Sōseki* (Tokyo, Raifusha, 1956).
11. Etō Jun, *Sōseki to Sono Jidai* (Tokyo, Shinchōsha, 1970).
12. Edwin McClellan, *Two Japanese Novelists: Sōseki and Tōson* (Chicago, Chicago University Press, 1969).

Synopses and Text

(N.B. The following are translator's notes unless otherwise indicated.)

1. *Edokko,* 江戸ッ子. *Edo,* 江戸, is the name of Tokyo prior to its being made the capital in 1868. The term *edokko* is now used to refer to a native Tokyoite but, strictly speaking, only those residents whose ancestors on both sides of the family have lived in the city for three generations or more qualify as true sons of *Edo.* Ideally, an *edokko* should have been born near the Zōjōji Temple in Shiba and raised in the shadow of the Myōjin Shrine in Kanda, i.e., *Shiba de umarete Kanda de sodatsu,* 芝で生まれて神田で育だつ. He is characterized by the traditional saying *yoigoshi no zeni wo motanu,* 宵越の銭は持たぬ. He does "not let the morrow's sun rise on today's earnings," i.e., he does not hesitate to spend his money in a belief that the future will provide. He is also characterized by the saying *satsuki no koi no fukinagashi,* 皐月の鯉の吹流し, i.e., "as empty as a carp pennant blowing in the May breezes." This is interpreted to mean that, in event of an argument or altercation, he does not hold grudges.
2. *Hatamoto,* 旗本. Literally, "bannermen." A top-ranking class of samurai warriors under the Tokugawa shogunate. These samurai served as hereditary, personal retainers to the Shogun and were known for their loyality and honesty.
3. *Obā-san ko,* おばあさん子.
4. *Kahogoji,* 過保護児.
5. *On ni kiru,* 恩に着る. Readers of Ruth Benedict's early sociological study of Japanese culture, *The Chrysanthemum and the Sword,*

should find this passage of interest. Also cf. Doi Takeo, *The Anatomy of Dependence,* trans. John Bester (Tokyo, Kōdansha International, 1972), pp. 87-91ff.

6. Note that the term "higher school" is used in this translation to denote *kōtōgakkō,* 高等学校, which is most closely equivalent to the "gymnasium" of the German educational system. "College" or "university" is used to translate *daigaku,* 大学. Also note that the academic year in Sōseki's day began in early September rather than in April as it presently does in Japan.
7. *Giri,* 義理.
8. *Ninjō,* 人情.
9. Erik H. Erikson, "On the Sense of Human Identity," *Psychoanalytic Psychiatry and Psychology,* ed. Robert Knight and Cyrus Friedman (New York, International University Press, 1954), pp. 351-364.
10. *Ankonshasu hipokurishii* アンコンシャス・ヒポクリシイ. A term coined in a review by Sōseki to describe the behavior of Felicitas, heroine of Hermann Sudermann's novel *Es War*. He further indicates that he was attempting to create a similar character in writing *Sanshirō*. Cf. Komiya Toyotaka, *Sōseki no Geijutsu* (Tokyo, Iwanami Shoten, 1942), pp. 152-160ff.
11. English dramatist Thomas Southerne, 1660-1746. The line appears in his drama, *Oroonoko*, written in 1696 and adapted from a novel of the same name by Aphra Behn.
12. *Kōtōyūmin* 高等遊民. This term was invented by Sōseki to describe characters such as Daisuke of *Sorekara* and Sunaga and Matsumoto of *Higan sugi made* who are well-educated and lead lives of leisure.
13. The characters in question are *kin*, 近, and *kon*, 今, used in the sense of "lately" (as in *kinrai*, 近来) and "today" (*konnichi*, 今日) respectively. They would be familiar to even a first-grader. It is typical of Sōseki's artistic skill and psychological insight that he should have Sōsuke, a man obsessed with his past, stumble over words relating to the present.
14. In a brief introduction to the novel, Sōseki states his hope that the work will be found both entertaining and true to his unique style of writing. "I hold the belief that I am myself. I have no intention, therefore, of troubling with what 'ism' I am. I am neither naturalist, symbolist, nor romanticist with a 'neo.'" Natsume Sōseki, *Sōseki Zenshū* (Tokyo, Iwanami Shoten, 1966), V, 6.
15. Chitani Shichirō, *Sōseki no Byōseki* (Tokyo, Keisō Shobo, 1963).

16. In recent years Chitani's psychiatric diagnosis of Ichirō's case has enjoyed considerable prominence. As I have stated, however, the schizophrenic pattern is more conspicuous. Relevant to this point is H's remark that Ichirō "was born tall and thin, while I was short and heavyset." According to Ernst Kretschmer's *Körperbau und Charakter,* schizophrenia is more frequent among slender types and depression among heavy ones. Kretschmer's book was not published until 1921. Considering that *Kōjin* appeared in 1913, one is again struck by the keenness of Sōseki's powers of observation. (Author's note)
17. *Ayasu,* あやす.
18. *Sensei,* 先生. *Sensei* is not really a name but a term of address that is closest in meaning and usage to the French *maître.*
19. *Oyaji,* 親父.
20. *Ofukuro,* おふくろ.
21. *Michikusa,* 道草. Literally, "grass alongside a road." When used idiomatically as in *michikusa wo kuu* 道草を食う, it refers to the way a horse will stop on a journey to indulge its taste for wayside grasses. Hence, "to loiter or to tarry."
22. In a letter of November 15, 1916 to Tomizawa Keidō, Sōseki writes, "I know this may sound strange, but I am the sort of fool who, at the age of fifty, decides for the first time in his life to take up the Way. When I consider how far I am from fulfillment, I am aghast at the greatness of the distance." (Author's note)

INDEX

HARVARD EAST ASIAN MONOGRAPHS

1. Liang Fang-chung, *The Single-Whip Method of Taxation in China.* 1956. 79 pp.
2. Harold C. Hinton, *The Grain Tribute System of China, 1845-1911.* 1956. 171 pp.
3. Ellsworth C. Carlson, *The Kaiping Mines, 1877-1912.* 1971. 235 pp.
4. Chao Kuo-chün, *Agrarian Policies of Mainland China: A Documentary Study, 1949-1956.* 1957. 290 pp.
5. Edgar Snow, *Random Notes on Red China, 1936-1945.* 1957. 164 pp.
6. Edwin George Beal, Jr., *The Origin of Likin, 1835-1864.* 1958. 204 pp.
7. Chao Kuo-chün, *Economic Planning and Organization in Mainland China: A Documentary Study, 1949-1957.* Vol. I: 275 pp. 1959. Vol. II: 280 pp. 1960.
8. John K. Fairbank, *Ch'ing Documents: An Introductory Syllabus.* Vol. I: 141 pp. Vol. II: 41 pp. 1965.
9. Helen Yin and Yi-chang Yin, *Economic Statistics of Mainland China, 1949-1957.* 1960. 110 pp.
10. Wolfgang Franke, *The Reform and Abolition of the Traditional Chinese Examination System.* 1960. 110 pp.
11. Albert Feuerwerker and S. Cheng, *Chinese Communist Studies of Modern Chinese History.* 1961. 313 pp.
12. C. John Stanley, *Late Ch'ing Finance: Hu Kuang-yung as an Innovator.* 1961. 313 pp.
13. S. M. Meng, *The Tsungli Yamen: Its Organization and Functions.* 1962. 152 pp.
14. Ssu-yü Teng, *Historiography of the Taiping Rebellion.* 1962. 188 pp.
15. Chun-Jo Liu, *Controversies in Modern Chinese Intellectual History: An Analytic Bibliography of Periodical Articles,*

Mainly of the May Fourth and Post-May Fourth Era. 1964. 215 pp.

16. Edward J. M. Rhoads, *The Chinese Red Army, 1927-1963: An Annotated Bibliography.* 1964. 202 pp.
17. Andrew J. Nathan, *A History of the China International Famine Relief Commission.* 1965. 114 pp.
18. Frank H. H. King (ed.) and Prescott Clarke, *A Research Guide to China-Coast Newspapers, 1822-1911.* 1965. 245 pp.
19. Ellis Joffe, *Party and Army: Professionalism and Political Control in the Chinese Officer Corps, 1949-1964.* 1965. 210 pp.
20. Toshio G. Tsukahira, *Feudal Control in Tokugawa Japan: The Sankin Kōtai System.* 1966. 240 pp.
21. Kwang-Ching Liu, ed., *American Missionaries in China: Papers from Harvard Seminars.* 1966. 316 pp.
22. George Moseley, *A Sino-Soviet Cultural Frontier: The Ili Kazakh Autonomous Chou.* 1966. 171 pp.
23. Carl F. Nathan, *Plague Prevention and Politics in Manchuria, 1910-1931.* 1967. 112 pp.
24. Adrian Arthur Bennett, *John Fryer: The Introduction of Western Science and Technology into Nineteenth-Century China.* 1967. 169 pp.
25. Donald J. Friedman, *The Road from Isolation: The Campaign of the American Committee for Non-Participation in Japanese Aggression, 1938-1941.* 1968. 132 pp.
26. Edward Le Fevour, *Western Enterprise in Late Ch'ing China: A Selective Survey of Jardine, Matheson and Company's Operations, 1842-1895.* 1968. 223 pp.
27. Charles Neuhauser, *Third World Politics: China and the Afro-Asian People's Solidarity Organization, 1957-1967.* 1968. 107 pp.
28. Kungtu C. Sun, assisted by Ralph W. Huenemann, *The Economic Development of Manchuria in the First Half of the Twentieth Century.* 1969. 134 pp.

29. Shahid Javed Burki, *A Study of Chinese Communes, 1965.* 1969. 117 pp.

30. John Carter Vincent, *The Extraterritorial System in China: Final Phase.* 1970. 134 pp.

31. Madeleine Chi, *China Diplomacy, 1914-1918.* 1970. 213 pp.

32. Clifton Jackson Phillips, *Protestant America and the Pagan World: The First Half Century of the American Board of Commissioners for Foreign Missions, 1810-1860.* 1969. 380 pp.

33. James Pusey, *Wu Han: Attacking the Present through the Past.* 1970. 94 pp.

34. Ying-wan Cheng, *Postal Communication in China and Its Modernization, 1860-1896.* 1970. 150 pp.

35. Tuvia Blumenthal, *Saving in Postwar Japan.* 1970. 120 pp.

36. Peter Frost, *The Bakumatsu Currency Crisis.* 1970. 87 pp.

37. Stephen C. Lockwood, *Augustine Heard and Company, 1858-1862: American Merchants in China.* 1970. 125 pp.

38. Robert R. Campbell, *James Duncan Campbell: A Memoir by His Son.* 1970. 145 pp.

39. Jerome Alan Cohen, ed., *The Dynamics of China's Foreign Relations.* 1970. 139 pp.

40. V. V. Vishnyakova-Akimova, *Two Years in Revolutionary China, 1925-1927,* tr. Steven I. Levine. 1971. 345 pp.

41. Meron Medzini, *French Policy in Japan during the Closing Years of the Tokugawa Regime.* 1971. 267 pp.

42. *The Cultural Revolution in the Provinces.* 1971. 267 pp.

43. Sidney A. Forsythe, *An American Missionary Community in China, 1895-1905.* 1971. 152 pp.

44. Benjamin I. Schwartz, ed., *Reflections on the May Fourth Movement: A Symposium.* 1972. 140 pp.

45. Ching Young Choe, *The Rule of the Taewŏn'gun, 1865-1873: Restoration in Yi Korea.* 1972. 287 pp.

46. W. P. J. Hall, *A Bibliographical Guide to Japanese Research on the Chinese Economy, 1958-1970.* 1972. 113 pp.

47. Jack J. Gerson, *Horatio Nelson Lay and Sino-British Relations, 1854-1864.* 1972. 350 pp.

48. Paul Richard Bohr, *Famine and the Missionary: Timothy Richard as Relief Administrator and Advocate of National Reform.* 1972. 301 pp.

49. Endymion Wilkinson, *The History of Imperial China: A Research Guide.* 1973. 234 pp.

50. Britten Dean, *China and Great Britain: The Diplomacy of Commercial Relations, 1860-1864.* 1974. 223 pp.

51. Ellsworth C. Carlson, *The Foochow Missionaries, 1847-1880.* 1974. 259 pp.

52. Yeh-chien Wang, *An Estimate of the Land-Tax Collection in China, 1753 and 1908.* 1973. 192 pp.

53. Richard M. Pfeffer, *Understanding Business Contracts in China, 1949-1963.* 1973. 147 pp.

54. Han-sheng Chuan and Richard Kraus, *Mid-Ch'ing Rice Markets and Trade, An Essay in Price History.* 1975. 238 pp.

55. Ranbir Vohra, *Lao She and the Chinese Revolution.* 1974. 199 pp.

56. Liang-lin Hsiao, *China's Foreign Trade Statistics, 1864-1949.* 1974. 297 pp.

57. Lee-hsia Hsu Ting, *Government Control of the Press in Modern China, 1900-1949.* 1974. 318 pp.

58. Edward W. Wagner, *The Literati Purges: Political Conflict in Early Yi Korea.* 1974. 238 pp.

59. Joungwon A. Kim, *Divided Korea: The Politics of Development, 1945-1972.* 1975. 445 pp.

60. Noriko Kamachi, John K. Fairbank, and Chuzo Ichiko, *Japanese Studies of Modern China Since 1953: A Bibliographical Guide to Historical and Social-Science Research on the Nineteenth and Twentieth Centuries, Supplementary Volume for 1953-1969.* 1975. 610 pp.

61. Donald A. Gibbs and Yun-chen Li, *A Bibliography of Studies and Translations of Modern Chinese Literature, 1918-1942.* 1975. 239 pp.

62. Robert H. Silin, *Leadership and Values: The Organization of Large-Scale Taiwanese Enterprises*. 1975.

63. David Pong, *A Critical Guide to the Kwangtung Provincial Archives Deposited at the Public Record Office of London*. 1975. 203 pp.

64. Fred W. Drake, *China Charts the World: Hsu Chi-yü and His Geography of 1848*. 1975. 272 pp.

65. William A. Brown and Urgunge Onon, translators and annotators, *History of the Mongolian People's Republic*. 1976. 910 pp.

66. Edward L. Farmer, *Early Ming Government: The Evolution of Dual Capitals*. 1976.

67. Ralph C. Croizier, *Koxinga and Chinese Nationalism: History, Myth, and The Hero*. 1976.

68. William Jefferson Tyler, translator, *The Psychological World of Natsume Sōseki* by Dr. Takeo Doi. 1976.